PRAISE FO[illegible]

"Joe Schwarcz does it again with [illegible] .med exploration of the hot topics in science we've been bombarded with over the past few years! From the biology of vaccines (including the new mRNA variety) and immune boosting (spoiler: you can't) to the history of epidemiology and toilet paper, Schwarcz gives us the fact-filled low-down. In a world filled with misinformation and twisted science, this is a must-read!"
— Timothy Caulfield, Canada Research Chair in Health Law and Policy, bestselling author of *Is Gwyneth Paltrow Wrong About Everything?*

PRAISE FOR *A GRAIN OF SALT*

"Schwarcz's light touches of humor make the scientific information feel accessible and ensure that it's entertaining. With enough facts to soothe anxious, health-conscious individuals as well as some good tidbits to share, this enlightening collection offers every reader something new to learn and marvel over."
— *Booklist*

PRAISE FOR *A FEAST OF SCIENCE*

"Huzzah! Dr. Joe does it again! Another masterwork of demarcating non-science from science and more generally nonsense from sense. The world needs his discernment."
— Dr. Brian Alters, Professor, Chapman University

PRAISE FOR *THE FLY IN THE OINTMENT*

"Joe Schwarcz has done it again. In fact, he has outdone it. This book is every bit as entertaining, informative, and authoritative as his previous celebrated collections, but contains enriched social fiber and 10 percent more attitude per chapter. Whether he's assessing the legacy of Rachel Carson, coping with penile underachievement in alligators, or revealing the curdling secrets of cheese, Schwarcz never fails to fascinate."
— Curt Supplee, former science editor, *Washington Post*

PRAISE FOR *DR. JOE AND WHAT YOU DIDN'T KNOW*

"Any science writer can come up with the answers. But only Dr. Joe can turn the world's most fascinating questions into a compelling journey through the great scientific mysteries of everyday life. *Dr. Joe and What You Didn't Know* proves yet again that all great science springs from the curiosity of asking the simple question . . . and that Dr. Joe is one of the great science storytellers with both all the questions and answers."
— Paul Lewis, president and general manager, Discovery Channel

PRAISE FOR *THAT'S THE WAY THE COOKIE CRUMBLES*

"Schwarcz explains science in such a calm, compelling manner, you can't help but heed his words. How else to explain why I'm now stir-frying cabbage for dinner and seeing its cruciferous cousins — broccoli, cauliflower, and brussels sprouts — in a delicious new light?"
— Cynthia David, *Toronto Star*

PRAISE FOR *RADAR, HULA HOOPS, AND PLAYFUL PIGS*

"It is hard to believe that anyone could be drawn to such a dull and smelly subject as chemistry — until, that is, one picks up Joe Schwarcz's book and is reminded that with every breath and feeling one is experiencing chemistry. Falling in love, we all know, is a matter of the right chemistry. Schwarcz gets his chemistry right, and hooks his readers."
— John C. Polanyi, Nobel Laureate

Also by Dr. Joe Schwarcz

Science Goes Viral: Captivating Accounts of Science in Everyday Life

A Grain of Salt: The Science and Pseudoscience of What We Eat

A Feast of Science: Intriguing Morsels from the Science of Everyday Life

Monkeys, Myths, and Molecules: Separating Fact from Fiction, and the Science of Everyday Life

Is That a Fact?: Frauds, Quacks, and the Real Science of Everyday Life

The Right Chemistry: 108 Enlightening, Nutritious, Health-Conscious and Occasionally Bizarre Inquiries into the Science of Everyday Life

Dr. Joe's Health Lab: 164 Amazing Insights into the Science of Medicine, Nutrition and Well-Being

Dr. Joe's Brain Sparks: 179 Inspiring and Enlightening Inquiries into the Science of Everyday Life

Dr. Joe's Science, Sense and Nonsense: 61 Nourishing, Healthy, Bunk-Free Commentaries on the Chemistry That Affects Us All

Brain Fuel: 199 Mind-Expanding Inquiries into the Science of Everyday Life

An Apple a Day: The Myths, Misconceptions and Truths About the Foods We Eat

Let Them Eat Flax: 70 All-New Commentaries on the Science of Everyday Food & Life

The Fly in the Ointment: 70 Fascinating Commentaries on the Science of Everyday Life

Dr. Joe and What You Didn't Know: 177 Fascinating Questions and Answers About the Chemistry of Everyday Life

That's the Way the Cookie Crumbles: 62 All-New Commentaries on the Fascinating Chemistry of Everyday Life

The Genie in the Bottle: 64 All-New Commentaries on the Fascinating Chemistry of Everyday Life

Radar, Hula Hoops, and Playful Pigs: 67 Digestible Commentaries on the Fascinating Chemistry of Everyday Life

QUACK QUACK

The Threat of Pseudoscience

DR. JOE SCHWARCZ

This book is also available as a Global Certified Accessible™ (GCA) ebook. ECW Press's ebooks are screen reader friendly and are built to meet the needs of those who are unable to read standard print due to blindness, low vision, dyslexia, or a physical disability.

Purchase the print edition
and receive the ebook free.
For details, go to ecwpress.com/ebook.

Published by ECW Press
665 Gerrard Street East
Toronto, Ontario, Canada M4M 1Y2
416-694-3348 / info@ecwpress.com

Cover design: David A. Gee

LIBRARY AND ARCHIVES CANADA CATALOGUING IN PUBLICATION

Title: Quack quack : the threat of pseudoscience / Dr. Joe Schwarcz.

Names: Schwarcz, Joe, author.

Identifiers: Canadiana (print) 20220183627 | Canadiana (ebook) 20220183678

ISBN 978-1-77041-658-1 (softcover)
ISBN 978-1-77852-023-5 (ePub)
ISBN 978-1-77852-024-2 (PDF)
ISBN 978-1-77852-025-9 (Kindle)

Subjects: LCSH: Quacks and quackery. | LCSH: Pseudoscience. | LCSH: Common fallacies.

Classification: LCC R730 .S39 2022

| DDC 615.8/56—dc23

This book is funded in part by the Government of Canada. *Ce livre est financé en partie par le gouvernement du Canada.* We also acknowledge the support of the Government of Ontario through Ontario Creates.

Canada

PRINTED AND BOUND IN CANADA

PRINTING: MARQUIS 5 4 3 2 1

CONTENTS

INTRODUCTION

I've often been asked about who sparked my enthusiasm, some would say obsession, for separating sense from nonsense. That credit goes to three individuals, two real and one fictional: Harry Houdini with his exposés of the antics of mediums, James Randi with his tireless efforts to unmask charlatans and promote critical thinking, and Sherlock Holmes with his emphasis on coming to conclusions only if they are supported by facts.

I was introduced to magic at a young age by a performer at a birthday party and was intrigued enough to start reading about the subject. You do not have to delve deeply into the field before encountering Houdini, a man whose name to this day is virtually synonymous with magic. While I was taken with his exploits on the stage, my attention was also drawn to Houdini's crusade against charlatans who were using conjuring tricks to convince the gullible that they were communicating with the spirit world. The extent to which some people would go to fool others was an eye-opener for me. And very disturbing.

I became annoyed with claims of bending spoons with the power of the mind, "psychic surgeons" removing tissues without an incision, and psychics moving objects through "psychokinesis" when it was clear that all these were performed by standard magic effects. Then when I began to pursue science I encountered a different kind of huckster. One who would use sketchy or outright false concepts to promote products or treatments either out of ignorance or for financial gain. These are the homeopaths, the "alkaline water" advocates, and the most reprehensible, the quack cancer-cure claimants. There is yet another class of mountebank: highly educated people who have managed to garner publicity for having latched on to some eccentric idea and keep coming up with more and more outrageous arguments in an attempt to prove their point and stay in the limelight. Often they are delusional, but there are also cases of researchers who are so wedded to their ideas that they will not change their mind even in the face of contrary evidence.

Reading Houdini's life story introduced me to Sir Arthur Conan Doyle, the creator of Sherlock Holmes. The magician met Conan Doyle in 1920 when he was touring Britain and the two struck up one of the strangest friendships ever. Houdini was the scourge of the spiritualists and Conan Doyle was an ardent believer in the possibility of contact with the departed. That was a real enigma for Houdini, since Sherlock Holmes was totally scientifically minded, yet his creator believed in spirits! Curiously, in "The Adventure of the Sussex Vampire," Holmes firmly states: "This agency stands flat-footed upon the ground, and there it must remain. The world is big enough for us. No ghosts need apply." Amazingly, Conan Doyle was convinced that Houdini had supernatural powers since he could not explain how he performed his magical feats. This placed the magician in a difficult position because, abiding by the code of magicians, he would not reveal his secrets.

Having become acquainted with the bizarre Conan Doyle–Houdini friendship, I was stimulated to start reading the Sherlock Holmes stories and became enthralled with the detective's emphasis on making scientific observations and "cause and effect" conclusions. The first story I read was "A Scandal in Bohemia," where I would come across a Holmes quote that I would reference throughout my life: "It is a capital mistake to theorize before one has data. Insensibly one begins to twist facts to suit theories, instead of theories to suit facts." Throughout my career, much to my dismay, I discovered that not everyone abides by this dictum. And my archenemy, pseudoscience, rears its ugly head when facts are tortured until they fit some pet theory.

Over the last forty-plus decades I've encountered numerous claims of paranormal abilities, hoaxes, miraculous cancer cures, and devices to diagnose or treat disease, all of which fall under the pseudoscience umbrella. In this collection, I will try to provide a taste of the widespread nonsensical beliefs I have encountered and hope to convince you of the importance of separating sense from nonsense. Through my public lectures, radio shows, and various media appearances I have also come across a host of organizations and individuals that engage in dishonest practices and claim to have some special knowledge or skill that they do not actually possess. These are the "quacks." Let's get started.

QUACK QUACK

I collect ducks. Not live ones, but every other kind. I have wooden, ceramic, metal, plastic, and rubber ducks. I have ducks that waddle down an incline. I have windup ducks, inflatable ducks, stuffed ducks, and even a life-size wooden duck that performs a magic trick. Pick a card, any card, place it back in the deck, shuffle, and the duck will find it.

Why this infatuation with ducks? Because they quack. And quackery fascinates, amuses, and, above all, disturbs me. Of course, the kind of quackery I'm talking about doesn't refer to ducks, but to a special breed of ignorant pretenders to knowledge — usually, but not necessarily, in matters related to health. Quacks boast of providing cures that sound wondrous, but turn out to be no more than clever deceptions.

The origin of the term "quack" is somewhat murky, but may well have been inspired by the resemblance between the sound a duck makes and the rapid-fire oratory of charlatans plying their nostrums. Alternatively, the term may derive from the archaic Dutch expression "quacksalber," referring to itinerant mountebanks who hawked various "salves" that were promoted as having miraculous properties but fell well short of the claims.

The eighteenth and nineteenth centuries are often regarded as the golden age of quackery, but today's peddlers of sham don't have to take a back seat to any of the colorful schemers of previous eras. Aided and abetted by the Internet, they effectively turn gullibility and illness into healthy profits with a mishmash of pseudoscience and seductive testimonials.

The hallmarks of quackery include the bashing of conventional medicine, the use of words of praise from supposedly contented patients, and extravagant claims for painless cures for virtually all diseases. For example, "Dr." Brodum's Nervous Cordial and Botanical Syrup, introduced in 1801, was good for "excruciating rheumatic pains and contraction of the joints," as well as for curing the "indiscretions of youth." Questionable qualifications are also often part of the package. Brodum's bogus medical degree was from the Marischal University of Aberdeen, easily confused with the legitimate University of Aberdeen.

Potter's Vegetable Catholicon cured diseases of the liver and "debility resulting from intemperance and dissipation." General Augustus J. Pleasonton promoted the idea that light rays filtered through a special blue glass arrest disease and restore health. Then there was "Dr." Edrehi's Amulet containing a berry that released a scent "preventative of fevers and general decline of the system." As a bonus, the berry's strong odor would protect clothing from moths. Quacks also sold intriguing devices, such as Dr. Hawley's mechanical treatment for impotence, cleverly named the Erector.

While most quack remedies were devoid of biologically active ingredients, James Morison's Vegetable Universal Medicine contained aloes, jalap, gamboges, colocynth, and rhubarb, all of which are plants with laxative properties. Morison invented the Hygeian System, based on the notion that all pain and disease arise from impurities in the blood and the only effective way of eliminating these impurities is with vegetable purgatives. Calling himself "The Hygeist," he tirelessly attacked the medical profession as the enemy, claiming that "the old medical science is completely wrong," all the while ignoring the potentially lethal effects of the large number of pills he recommended, which of course were only available from his agents.

Scientific advances often fostered the promotion of quack cures. William Radam used the discoveries of Robert Koch and Louis Pasteur to promote his Microbe Killer, which was nothing other than a useless dilute solution of hydrochloric and sulphuric acids. The introduction of electricity launched a variety of electro-galvanic belts that were to be worn around the middle to cure "nervous debility, female complaints, catarrh and diseases of the blood." The belts contained copper and zinc disks that constituted a battery capable of producing a mild burning sensation to indicate to the wearer that something was happening. Something was indeed happening, but it had nothing to do with curing any disease. And then there was the Health Jolting Chair, which targeted the fairer sex, "members of which are particularly prone to neglect the taking of proper exercise and consequently are robbed of the sweet pure breath, vigorous mental action and vivacious manner characteristic of healthy young womanhood." The chair was equipped with machinery that would provide the jolts that "exercised the internal organs of the body necessary to health."

So that was then. How about now? Well, you can buy a quantum balance crystal that will restore the light frequencies missing within our quantum energy field, a result of the "noxious energy beam that emanates from digital TVs." You can also invest in the Photon Genie, which can "devitalize pathogens and detoxify the body with nourishing photobionic energy effectively delivered by both an ionized noble gas energy transmission and deeply penetrating mega frequency life-force energy waves." The babble may have been updated, but nonsense is nonsense even if it is cloaked in the garb of science.

And did you know that pine pollen can "elevate sexual libido [is there any other kind?], increase fertility and decrease the symptoms of aging"? Or that wearing tourmaline infrared ray socks can improve blood circulation, increase metabolism and enhance general health? How about fucoidan? This long-chain carbohydrate isolated from seaweed makes "cancer cells self destruct in as little as seventy-two hours, so cancer cells die by the thousands while healthy cells remain untouched." Of course this is all hushed up by an evil pharmaceutical industry, as is the "fact" that an extract of the rainforest fruit graviola is 10,000 times as strong as a common chemotherapy drug.

For more mind-numbing claptrap, how about "all natural liquid oxygen" drops under the tongue to fight jet lag, fatigue, hangover, and aging skin? And while at it, you might want to remove the worm-infested toxic sludge from your bowels, a consequence of a "diet filled with food additives, pesticides and other chemicals." All you need is the breakthrough "100% natural pill" that "can put your doctor out of business." Is this really that different from Ching's Patent Worm Lozenges, advertised in 1802 as of "peculiar importance to those afflicted with internal complaints?" It seems human credulity is a constant, unaffected by the march of science. Now you see why I collect ducks?

PSEUDOSCIENCE

In 1964 the U.S. Supreme Court was stymied in its attempts to come up with a definition of obscenity. That's when Justice Potter Stewart rendered his opinion with the famous phrase "I know it when I see it." Today we face a similar issue with defining "pseudoscience" as this plague encroaches on science with more and more vigor. A definition is hard to come by because pseudoscience takes on so many forms. Most scientists, though, would agree that when they see it, they know it. But this is not necessarily the case for people not well versed in science. Unlike obscenity, which in spite of being hard to define is readily recognizable, pseudoscience can masquerade as science and wreak intellectual and physical havoc.

In essence, pseudoscience encompasses any belief, process, or claim that pretends to have a scientific basis but actually has none. While real science accumulates facts and formulates testable theories to gain an understanding of the physical world, pseudoscience relies on anecdotes, ideology, and cherry-picked data to support preconceived notions. Conventional science is a self-correcting, continuously evolving process based on critical thinking and plausible theories supported by peer-reviewed research, whereas pseudoscience is often mired in dogma that is resistant to change.

Homeopathy would be a typical example of a pseudoscience. It tries to rationalize its tenet that drugs become more potent as they are diluted and shaken by claiming that the process leaves a molecular imprint on water, and that this imprint has healing properties. Based on what we know about the properties and behavior of water, and we know a lot, such a claim has no scientific basis. Ditto for "distance healing." Here the proposition is that a healer can examine a photograph of a patient and diagnose energy blockages that are causing disease. They can then proceed to heal by transmitting some sort of consciousness waves that clear the blockages. The distance between the healer and the patient is irrelevant. Aside from the implausibility of diagnosing disease from a picture, science knows of no form of energy transmission that doesn't fall off with distance.

Proponents of distance healing often introduce pseudoscientific language that can sound very seductive. "The human body has a resonant frequency and coherence is its natural state." Or "chimps and humans have similar DNA, so that couldn't explain the difference between them, the explanation is the morphogenetic field that informs which parts of DNA the body will access for its development." Needless to say, healers can restore health by altering this "morphogenetic field," a totally fictional and meaningless term.

When scientists start raising eyebrows at such mindless twaddle, the pseudoscience champions unleash their usual attacks, claiming that scientists are closed-minded and can only think in terms of limited paradigms. They love to compare themselves to the Wright brothers or to Columbus, who were laughed at. Indeed the Wrights were ridiculed when they proposed to build a flying machine, and a Spanish Royal Commission did reject Columbus's proposal to sail west, saying that "so many centuries after the Creation, it is unlikely that anyone could find hitherto unknown lands of any value." But when Columbus landed in America and the Wright brothers flew at Kitty Hawk, the laughter died. It was killed by evidence! As Carl Sagan cleverly said, they may have laughed at Columbus and the Wright brothers, but they also laughed at Bozo the Clown.

Scientists are quite prepared to stop laughing when evidence is provided. Doctors poked fun at Dr. Barry Marshall for his claim that ulcers were caused by bacteria, but they immediately embraced the idea when Marshall showed that ulcers could be cured with antibiotics. Where is the comparable evidence for astrology, iridology, crystal healing, quantum healing, magnetic healing, "qi" channels, telepathy, or creationism? Where is the proof that John Kanzius's radio waves can cure cancer or that Hulda Clark's Zapper can zap it away? There isn't any, and Hulda is not going to produce any, given she passed away from cancer in 2009.

Today, aided and abetted by the Internet, pseudoscience is enjoying a golden age. Besides the intellectual muddling they cause, scientifically unsupported claims about therapeutic interventions that are not science-based offer false hope. Worse, they may steer patients away from proven conventional treatments.

We can discuss pseudoscience, we can try to define it, we can try to distinguish it from science, and we can try to alert people to some of its more troublesome features. But we can never eliminate it. As famed science writer Isaac Asimov opined, "Inspect every piece of pseudoscience and you will find a security blanket, a thumb to suck, a skirt to hold. What does the scientist have to offer in exchange? Uncertainty! Insecurity!" True enough, but we can also offer a scientific look at pseudoscience.

SNAKE OIL

It must have been quite a sight at the World's Exposition in Chicago in 1893. Clark Stanley, better known as "The Rattlesnake King," reached into a sack, plucked out a snake, slit it open, and plunged it into boiling water. When the fat rose to the top, he skimmed it off and used it on the spot to create Stanley's Snake Oil, a liniment that was immediately snapped up by the throng that had gathered to watch the spectacle. Little wonder. After all, Stanley had proclaimed that the liniment would cure rheumatism, neuralgia, sciatica, lumbago, sore throat, frostbite, and even toothache.

It wasn't too hard to convince the onlookers about the wonders of the liniment, particularly when it came to arthritis. All Stanley had to do was point out that snakes obviously did not suffer from this condition and seemed well lubricated internally. The crowd lapped up the hype and shelled out the money. And many claimed immediate relief from their pain! Could there have been something to this remedy? Maybe. But if it offered any relief, it wasn't due to any rattlesnake oil. It seems the snake act was only for show, and the liniment that was actually sold had been previously prepared. And not from snakes! Chemical analysis of a surviving sample revealed a mixture of mineral oil, beef fat, turpentine, camphor, and red pepper. As it turns out, both camphor and capsaicin, the latter found in red pepper, do have some painkilling effect when rubbed on aching joints. But the most effective ingredient in Stanley's Snake Oil was a good dose of placebo.

Actually, Clark Stanley didn't come up with the idea of snake oil as a remedy. That notion can be credited to the ancient Chinese who rubbed the oil on aching joints and claimed relief. Stanley probably heard about the remedy from Chinese immigrants who had come to America to seek their fortune. Many found jobs building the Transcontinental Railroad and could well have used snake oil they had brought along to help deal with the backbreaking work.

Chinese snake oil, though, was certainly not made from rattlesnakes. Traditionally, the oil was extracted from the fat sack of the Erabu sea snake. And that makes things interesting. As it turns out, sea snakes, like fish, are rich in omega-3 fats. Being cold-blooded animals, they have to be equipped with fats that don't harden in cold water, and omega-3 fats fit the bill. Erabu sea snakes actually are even richer in omega-3 fats than salmon, a popular source of these fats. We've heard a great deal about omega-3 fats in recent years, including their potential benefits in improving brain function, reducing the risk of heart attacks, alleviating depression, and even helping with arthritis.

Ahh, the arthritis. There really may be a connection here. Omega-3 fats are the body's precursors to certain prostaglandins that are known to have anti-inflammatory effects. So Chinese snake oil might actually have a beneficial effect. If you ingest it! But rattlesnake oil contains very little omega-3 fat, so even if Stanley's liniment had some rattlesnake oil, it wouldn't have been much use even if people swallowed it. But of course all Stanley asked them to swallow was the hype. This huckster may not have done much for his customers' health, but he did leave us with a legacy. Thanks to him, we commonly use the term "snake oil" for ineffective remedies. And some of today's snake oils make Stanley's product look respectable.

Wild Earth Animal Essences is a case in point. Just picture this scene. Daniel Mapel, a "spiritual psychologist," walks to a clearing in the Virginia woods and places a small bowl of water from a nearby stream on the ground. He then begins to walk in a large circle around the bowl, meditating and invoking the spirit of an animal. I don't quite understand how one attracts animal spirits, but apparently it involves tracing smaller and smaller circles while mentally asking the animal to share its gifts with humankind. By the time he reaches the center of the circle, Mapel claims, he and the animal are one. Whatever that means. At this time, he says, the energy of the animal is imparted to the bowl of water. This water is then formulated into essences that are sold as "vibrational remedies." I kid you not. Each of these, according to Mapel, contains the energetic imprint of the animal from which it is derived. He reassures us that no animals were captured or harmed to develop these products. Phew! It is a relief to know that no animal has noted the theft of its spirit.

Oh yes, along with the spirit-infused water, the essences also contain a small amount of brandy as a "vibrational preservative." Without the brandy, we are told, the vibration would quickly dissipate. Maybe without the brandy Mapel's ideas would also dissipate. But I digress. Let's get down to the essence of the Essence.

A customer can select from a wide array of these wonder products. There's eagle essence, for "soaring above earthly matters to gain perspective and clarity." But if what you want is "support in creating abundance at all levels of one's life," then you need rabbit essence. If you have a blood pressure issue, then buffalo essence may do the trick, at least according to one testimonial. A patient reports a remarkable drop in blood pressure from this essence, which is recommended "for slowing down and getting in touch with the resonance and rhythms of the Earth." And finally, if you need help for "facilitating initiation into the deepest, transpersonal realms of the psyche," then you need snake essence. The meaning of this claim is beyond me, but that is perhaps because I have not availed myself of the recommended five to seven drops of this concoction on a daily basis.

Maybe instead of spiritual vibrations, Mr. Mapel should consider putting some authentic snake oil into his snake essence. Why? Because in two recent studies, dietary sea snake oil enhanced maze-learning ability in mice. In other words, it made the mice smarter. Might be just the remedy our modern snake oil salesman needs to swallow.

SPIRITS OF SALT

Constantine Rhodocanaces had come to England sometime in the seventeenth century from the Greek isle of Chios and was brewing up medicine "in a Physicall Laboratory in London, next door to the Three Kings-Inne." I know this because it says so right on the cover of a booklet he published in 1663. A booklet that I now have! Not a copy, not a facsimile, but the original!

I have a penchant for unusual historical items, especially if they deal with chemistry or pseudoscience. My interest is aroused even more if these two coalesce. That is just the case for Rhodocanaces's publication about his wonder product, *Spirit of Salt of the World.* I came across a reference to this epic when I was doing some research for my book *A Grain of Salt,* which also deals with chemistry and pseudoscience. When I saw that *Spirit of Salt of the World* was available from an antique book dealer, I jumped, and I now am the proud owner of this wonderful little relic.

The first line that strikes your eye on the cover is "ALEXICACUS," which is Greek for "averter of evil." And that is just what Spirit of

Salt of the World promises to do. Avert all sorts of evil diseases from scurvy and "inflammation of the feet" to the "French Disease" and kidney stones. Rhodocanaces informs the reader that he works in a lab "where all manner of Chymicall preparations are carried on without any Sophistication or abuses whatsoever." This is where his Spirits of Salt of the World is "now Philosophically prepared and purified from all hurtful or Corroding Qualities, far beyond anything yet known to the World, being both safe and pleasant for the use of all Men, Women and Children." In those days, "sophistication" meant "deception" and "philosophically" was the term for our "scientifically." While there certainly was science involved in preparing the product, there was also deception.

So, what was Spirit of Salt of the World? "Spirit of salt" is an old term for hydrochloric acid. And that is what Rhodocanaces was peddling and claiming was a virtual panacea. Hydrochloric acid was well known at the time, having been discovered some eight hundred years earlier by the alchemist Jabir ibn Hayyan, who produced it by mixing salt with sulphuric acid. Hayyan is also credited for having discovered that distillation of a solution of a mineral we now know to be iron sulphate heptahydrate yields sulphuric acid. Rhodocanaces was not the first to claim therapeutic properties for hydrochloric acid, but he claims other products sold as Spirit of Saltare were outright dangerous when compared with his version. He tells us how he heard from the Right Worshipfull Thomas Middleton, apparently a trustworthy nobleman, that "a sick person making use of the common Spirit of Salt bought at the Apothecaries, died upon the taking of it." Having heard this, "charity towards my Neighbour commandeth me to make publick that hereafter greater caution may be had in using the vulgar Corrosive Spirit of Salt, instead whereof I make publick this, which is most innocent, and healthful, as may be seen in the following testimonies."

And there are testimonies galore. Mrs. Bird gave some of Rhodocanaces's acid to her children who were troubled with worms, "which it presently kill'd and brought away." A man "sick of an inveterate Head-ach, which afflicted him at certain times every day, having been left by Physicians, and in opinion near death, did after purgation once and again prescribed, make use of this Spirit of Salt for the space of a week and was thereby suddenly and strangely recovered." Strangely indeed.

There is more. Rhodocanaces's concoction "procures a good appetite and prevents putrifaction of anything in the Stomak, prevents Drunkennesse and sickness therefrom, expels diseases that arise from corrupted blood, by purifying it, that lay before idle, and settled in the veins, and makes it volatile, and to proceed more regularly in its circulation." It also "keeps arteries from all filth, or slime, and sends away the water that lurks betwixt the skin and flesh, by Stool and Urine."

Naturally, only the original will work. "There are some who pretend to make this Spirit according to my preparation, wherefore I think good to let the world know, that as yet this Secret hath not been communicated to any."

Why do I find this booklet so fascinating? Because it could have been written today. All you have to do is substitute Spirit of Salt with the name of one of the current supposed panaceas that entice the desperate and the worried-well with the same type of testimonials, smearing of competing products, and claims of secret breakthroughs. Of course, all those claims should be taken with "a grain of salt."

MEDICINE SHOWS

"All sold out, Doctor!" boomed the voice that would become familiar to theater audiences across America and Europe. But in 1897 Harry Houdini was a virtually unknown entertainer with the California Concert Company, a classic old-fashioned traveling medicine show. As the "Great Wizard," young Harry's job was to attract and hold an audience for "Doctor" Thomas Hill, the bewhiskered pitch doctor who would majestically stride onto the stage and lecture the audience on the virtues of the miraculous elixir he had developed. As the doctor spoke, Harry and his wife, Bess, circulated through the crowd, selling the potion. Periodically, Harry would bellow that the fantastic product was all sold out, but of course more bottles would soon magically appear. The crowd ate up the entertainment and drank in the elixir.

The latter years of the nineteenth century and the first two decades of the twentieth were the heyday of the medicine show. This blend of entertainment and hucksterism featured an array of acts centered on the appearance of a mountebank, usually attired in a top hat and frock coat, who would pitch the product. He would be addressed as "doctor" or "professor," although the only training he usually had was in the sheep industry: the pitch doctor had to be adept at pulling the wool over people's eyes before fleecing them. This was not hard to do, because people then, as indeed now, were eager to jump on simple solutions that promised treatments for complex health problems.

Many of the medicine shows had Asians, or at least actors masquerading as such, solemnly nodding onstage as the pitch man delivered his lecture. They had "brought the wisdom and healing secrets of the East," which they were now willing to share with backward Americans. But even more popular than eastern miracle workers were Indigenous people. People may have been suspicious of the native residents in other ways, but they fully believed that living in the woods had allowed them to master botanical medicine. Indigenous people in full regalia often led the parade as the medicine show entered a town, ready to splash a little color into dull and drab lives and, of course, to heal the sick.

Such was the case for perhaps the most popular of the shows, the Kickapoo Medicine Show, produced by the Kickapoo Indian Medicine Company. "Doc" Healy and "Texas Charlie" Bigelow had founded the company, which had no connection whatsoever to the Kickapoo Tribe of Oklahoma. They dreamed up Kickapoo Indian Salve for skin diseases, which featured the "best buffalo tallow" and was guaranteed not to contain hog's lard. They introduced Kickapoo Indian Worm Killer, which would expel the parasites they claimed caused so much human misery. And it did seem to do the job. People were shocked to see long stringy worms emerge from their bodies. Stringy they were, alright. The Worm Killer pills came equipped with their own worms. String had been wound into a tight ball and was packed cleverly into the pills, ready to unfold after ingestion.

It was, however, Kickapoo Indian Sagwa that was destined to become the main nostrum of the hundred or so Kickapoo Medicine Show companies that crisscrossed America. Healy and Bigelow claimed it was a virtual cure-all and even got "Buffalo Bill" Cody to endorse it in ads: "An Indian would as soon be without his horse, gun or blanket as without Sagwa." Indigenous people, of course, were always without Sagwa. Healy and Bigelow had invented the product and the name. Some of the advertising was more insidious. Then, as now, hucksters peddled their products by casting a shadow on medications prescribed by physicians. "Poisoned by Calomel — Cured by Sagwa," said an ad taking direct aim at a medication used at the time as a cathartic. Calomel (mercurous chloride) was not a great medication, to be sure, but there was no problem it caused that could be cured by Kickapoo Sagwa.

Most concoctions hawked at medicine shows were harmless and useless brews made from herbs, roots, and barks. They usually contained a hefty dose of alcohol, which of course increased the chance for satisfaction. Sometimes they even added a little opium to brighten the mood. A laxative was often included to demonstrate that toxins were being expelled.

Since most people do not realize that many ailments are self-limiting and are also unaware of the pitfalls of anecdotal evidence and the power of the placebo, medicine shows prospered. At least they did so until 1906, when the U.S. passed the Pure Food and Drug Act requiring manufacturers to list ingredients on patent-medicines and to curtail the hype surrounding these products.

With a toned-down pitch, and products that were no longer myster ious, the medicine shows began to lose their appeal. The fights with the law had also taken a toll and by 1914 even the Kickapoo road companies had buried the hatchet and disbanded. But medicine shows would be heard from again. And it would be science that would apparently legitimize them.

During the first half of the twentieth century, "vitamins" replaced Indian remedies as magical cures in the eyes of the public. Scientists had indeed shown that devastating diseases such as rickets, pellagra, scurvy, and beriberi responded to vitamin therapy, a finding that provided ammunition for hucksters to launch a barrage of wild, unsubstantiated claims. Dudley J. LeBlanc was a Louisiana state senator who founded the Happy Day Company, which would become famous for its star product, Hadacol. The "Happy Day" apparently was the day a doctor injected LeBlanc with some B vitamins for pains he had been having. He was so satisfied with the results that he developed Hadacol, a mixture of B vitamins and iron, to share with the world. LeBlanc was a master salesman. He published testimonials in newspapers in which people claimed benefits for arthritis, asthma, diabetes, epilepsy, heart trouble, tuberculosis, and ulcers. When the Food and Drug Administration got after him, he toned down the claims and cleverly suggested that Hadacol was good for what ailed you, as long as what ailed you was what Hadacol was good for.

In 1950, LeBlanc reinvented the traveling Medicine Show. A caravan of 130 vehicles toured the south, entertaining as many as 10,000 people a night who had come with Hadacol box tops as admission tickets. They were treated to musical classics like "The Hadacol Boogie" and "Who Put the Pep in Grandma?" Mickey Rooney, Chico Marx, and Burns and Allen performed comedy skits while clowns took long drinks from Hadacol bottles as their false eyes and noses lit up. There were amusing quips about Hadacol's powers. Like "have you heard about the ninety-five-year-old who was dying in the hospital? She was taking Hadacol but it didn't save her. Did save the baby though."

The fame of Hadacol spread and so did the profits. LeBlanc was spending an amazing million dollars a month on advertising and was grossing $20 million a year. The man even had a sense of humor. When on a talk show Groucho Marx asked LeBlanc what Hadacol was good for, he quickly replied that "it was good for five and a half million for me last year." By the early 1950s the public had apparently figured this out as well, and Hadacol was piled on that huge intellectual garbage heap of quack products.

Today, the traveling medicine show with its fascinating mix of fun and flimflam is gone. But not forgotten. If you get a chance, take in a Psychic Fair or a Health Food Expo at a hotel or convention hall and experience a throwback to the past. I did. There were crystal healers, astrologers, and dietary supplements galore. Some of the claims sounded like they came straight from the mouths of the Kickapoo pitch doctors. One booth even had a magician performing tricks to gather a crowd for a wondrous new nostrum. He was not very successful. His magic was roughly on par with that claimed for the product. He was certainly no Houdini.

MODERNIZING MOUNTEBANKS

The original term "mountebank" derives from the Italian "monta in banco," which literally means "getting up on a bench." So, mountebanks were sellers of dubious medicines who would mount on a bench and regale a gathering crowd with descriptions of their wondrous nostrums and elixirs that promised to restore health and endow men with unparalleled sexual powers. By the fifteenth century mountebanks were to be found on many a street corner in Europe, often accompanied by a "Merry Andrew," whose task was to attract an audience with an assortment of zany antics.

Historians suggest that the original Merry Andrew was actually Doctor Andrew Borde, physician to King Henry VIII, who was noted for his wit and captivating way of addressing the public on health matters. He produced merriment in his audiences and gave rise to imitators who may have lacked his knowledge but nevertheless managed to entertain the crowds with their buffoonery. These clowns came to be known as Merry Andrews, and by the seventeenth and eighteenth centuries no mountebank would be without his Merry Andrew.

The idea of blending comedy with medicine has historic origins. The famous anatomical theater at Bologna, where dissections were performed for medical students as early as the sixteenth century, featured a small door just above the lecturer's platform. When the professor found the students to be inattentive, he gave a signal, the door opened, and a fool's head would pop through. He cracked a joke and quickly withdrew. Students were roused from their somnolence, had a hearty laugh, and refocused their attention on the lecturer's words.

Once the Merry Andrew had attracted a crowd, the mountebank would do his best to liberate coins from pockets and purses. One of the most notorious quacks was Ben Willmore, whose spiel was actually recorded by an onlooker. "Behold this little vial, which contains in its narrow bounds what the whole universe cannot purchase, if sold to its true value. This admirable, this miraculous elixir, drawn from the hearts of Mandrakes, Phoenix livers, Tongues of Mermaids and distilled by contracted Sunbeams, has, besides the unknown virtue of curing all distempers both of mind and body, that divine one of animating the Heart of man to that degree, that however remiss, cold and cowardly by Nature, he shall become Vigorous and Brave. Gentlemen, if any of you present was at Death's Door, here's this, my Divine Elixir, will give you Life again." Wow!

Some mountebanks attacked physicians, much as is the case today. One Tom Jones was a classic example. He would rail against doctors whose only remedy for disease was to purge or bleed the patient. Of course he had the real solution. His Incomparable Balsam healed all sores, cuts and ulcers, his Specifick cured pain in a minute, and his

Pulvis Catharticus expelled poisons and fortified the heart against faintness. Actually, while probably useless, these nostrums were less likely to harm the patient than doctors' bleeding or purging.

Some of the mountebanks were more audacious in their challenge to physicians. John Pontaeus gained fame in the seventeenth century with Orvietan, his antidote to all poisons. He even offered proof. Physicians could administer a poison of their selection to his assistant, who would then be treated with a dose of Orvietan. The doctors accepted the challenge and decided on Aqua Fortis, or as we now know it, nitric acid. Not only was this known to be toxic, it was also highly corrosive. The quack's servant swallowed it, collapsed immediately, and was carried away, apparently dead. To the surprise of the physicians he reappeared the next day, none the worse for wear. It seems Pontaeus had a trick up his sleeve. Or more accurately, butter down his servant's throat. Before the "experiment" the assistant had swallowed a large dose of butter, enough to coat his mouth and throat, protecting him from the caustic liquid. After being carried off he was immediately given warm water and the water-butter mix made him so sick that he regurgitated the acid. So the story goes.

Pontaeus pulled other fast ones. He sold Green Salve, which supposedly healed all wounds, and he had an impressive demonstration to prove it. His assistant dipped his hands into molten lead, after which the apparently badly burned hands would be restored to perfect health with the magical salve. The audience happily anted up for the wonder product. Hopefully, they didn't try to reproduce the molten lead experiment because Pontaeus's "molten lead" was actually mercury that was dispensed with a ladle painted red to give the appearance of heat. The assistant's bloody hands displayed to the onlookers after being withdrawn from the "lead" were actually colored with vermilion (mercury sulphide) that had been hidden in his hand as he dipped them into the "molten" metal. The spectators were properly duped and Pontaeus's pocketbook swelled. As far as the assistant went, his occupational hazard was mercury poisoning.

There were other ingenious performances as well. In the early seventeenth century an Italian mountebank became famous for healing his arm with a miraculous oil after he had just gashed it with a knife. The healing oil was effective indeed, as long as the mountebank was equipped with a trick knife and had mastered the art of palming a piece of fabric soaked in chicken blood. Unfortunately, sometimes such effects backfired with the performer being accused of witchcraft. A young mountebank in Cologne was charged with witchcraft for having torn and restored a handkerchief in the presence of witnesses. What happened to the unfortunate performer is not known, but the magic trick has certainly survived. It's one of my favorites.

Today mountebanks have transformed themselves into what I propose to call "mountewebs." Instead of beguiling a few onlookers by mounting a bench, they snare multitudes by mounting websites. But their structured water, detox foot baths, ear candles, and energy bracelets are no more effective than Willmore's phoenix livers, mermaid tongues or contracted sunbeams. Plus ça change, plus c'est la même chose.

POKING INTO THE PUKE WEED DOCTOR

"I think we never had more need to be on our guard than at the present time. The people are crammed with poison drugs and the laws say they shall not examine and judge for themselves. The effects are pains, lingering sickness and death. Poison given to the sick by a person of the greatest skill will have exactly the same effect as it would if given by a fool."

You might think that quote comes from one of the numerous current websites that espouse the benefits of "natural treatments" over pharmaceutical drugs. It doesn't. It was actually uttered some two hundred years ago by Samuel Thomson, an uneducated pig farmer whose philosophy that any man could be his own physician took

America by storm in the nineteenth century. Eventually Thomson's self-help health care would be embraced by more than three million Americans and his ideas would even spread to Europe!

Thomson believed that all diseases could be cured by the use of herbs and heat. While his system of healing used some sixty herbs, *Lobelia inflata*, also known as puke weed or Indian tobacco, was front and center. Puke weed is a very appropriate name because ingesting the flowers, seeds, or roots of the plant makes people, let us just say, lose their breakfast. Thomson thought that before healing could commence, toxins had to be eliminated and puke weed was just right for the job. This was not a novel idea; conventional physicians at the time used mercurous chloride, better known as calomel, to purge patients. Lobelia's effects were less violent and Thomson's theory that people could cure themselves without relying on doctors appealed to a lot of people. Thomson was not the first to experiment with Lobelia. Native Americans treated dozens of ailments with the herb, ranging from fevers and venereal diseases to earaches and stiff necks. Lobelia also had a reputation as a love potion, which is hard to explain. Vomiting and love usually don't go together.

In Thomson's regimen, after the puke weed had finished its performance, it was time to restore the body's heat with steam baths and cayenne pepper, often in the form of an enema. If there were still complaints, other courses of treatment would follow with complex mixtures of herbs such as ginseng, peppermint, and horseradish, often mixed with camphor and turpentine.

As is often the case for "alternative therapies," Thomsonism was rooted in its patriarch's personal experience. Young Sam had become curious about a plant that grew wildly in his father's fields and for some strange reason tried chewing its pods. The effect was dramatic. It seems the man whom skeptics would eventually call "the puke doctor" had a funny bone. He convinced some of his friends to sample Lobelia and had a good laugh at their expense. His interest in plants aroused, Thomson began to follow the healing abilities of an "old wife" in the area who had a reputation for curing people with herbs, often consumed as a brew in hot water to produce sweating. He was intrigued when she managed to cure his rash with an herbal concoction. And then came a couple of catalytic events.

At the age of nineteen Thomson sustained an ankle injury that defied conventional treatment but resolved when he ingested comfrey root and applied a turpentine plaster. Two years later his mother contracted measles, which turned into what doctors called "galloping consumption." Thomson later commented that this had been an appropriate name because the doctors were riders who managed to gallop her out of the world in about nine weeks. But when he contracted the disease, he claimed to have cured himself with herbs. When Thomson later saw his wife cured by herbalists after doctors had failed, and he himself managed to cure his infant daughter of some skin condition by holding her over steaming water, Thomsonism was ready to gallop. Doctors, or educated quacks as Thomson called them, may have had their fancy degrees, but their blistering, bleeding, and purging were worse than useless. He could cure people with herbs and steam! Herbs grew towards the sun, the life-giving source of heat, and therefore must refresh one's health, the puke doc maintained.

As one might expect, physicians didn't take kindly to Thomson's attacks, and in 1809 one actually managed to accuse him of killing a patient with an overdose of Lobelia. The puke doctor had to await his trial in a cell for six weeks. At the trial he claimed that he had actually cured the patient, who was responsible for his own demise by venturing out into the cold instead of recuperating in a warm house. Meanwhile the prosecution claimed that the victim had succumbed because of excessive vomiting brought on by Lobelia. It is unlikely that this was the case, because Lobelia does not induce such dangerous vomiting, but Thomson actually was exonerated because of a botanical error by the prosecution. An astute defense attorney noted that the plant the prosecution had introduced as evidence was actually marsh rosemary and not Lobelia. That was enough for the case to be dismissed.

Thomsonians regarded the dismissal as vindication of their efforts and the movement continued to pick up steam. Indeed, it was the popularity of Thomsonism that led to the repeal of the laws that a number of states had passed restricting the practice of unconventional medicine. Opponents had labeled these "Black Laws" in reference to those that restricted Black Americans from practicing medicine. Eventually Thomson's movement faded when some of his followers grew tired of his attacks on physicians and his drive to end physician licensing. They wanted more legitimacy and urged more training and even the establishment of Thomsonian hospitals. That never happened, but Thomsonism holds a unique place in history as a pivotal factor in allowing unconventional treatments to legally flourish in spite of a lack of evidence for efficacy. It is one of the pillars upon which modern naturopathy rests.

Some practitioners today still recommend Lobelia as a "blood cleanser" and as a respiratory stimulant to treat asthma, while Boiron Laboratories markets *Lobelia inflata* as a homeopathic remedy to help people wean themselves off smoking. Unlike with the herbal preparation, there is no concern here about side effects since the homeopathic remedy has been diluted to the extent that it contains essentially no Lobelia. While Thomsonism as such has been relegated to the history books, the puke doctor's legacy of eschewing schooling and science in favor of reliance on "intuitive wisdom" and "nature's pharmacy" is unfortunately still with us.

THE CHEW-CHEW AND DO-DO MAN

In the late 1890s, a number of scientists around the U.S. received a strange package in the mail. The contents, labeled "economic ash," were actually samples of human excrement. The sender was Horace Fletcher, one of the most famous dietary gurus of the era. The motive? To show scientists that proper digestion produced stools "with no stench, no evidence of putrid bacterial decomposition, only with the odor of warm earth or a hot biscuit." And what was the secret of proper digestion? To eat like Horace Fletcher or, as said one of his famous devotees, Dr. John Harvey Kellogg, to "Fletcherize."

Horace Fletcher was born in 1849 and spent his youth traveling around the world. By the time he turned forty he had become a rich man through importing Japanese art, toys, and novelties. His portly appearance spoke of the good life. And then came a pivotal moment. Fletcher applied for life insurance but was turned down on the basis of an unfavorable medical exam. Now he became obsessed with health. The man who up to now had shown no interest in matters of science began to expound a theory of health. "Troubles come from too much of many things," he said, "among them too much food and too much worry."

Worry was to be countered through the process of "menticulture." Today we would call it stress management. But the nutritional causes of disease were to be eliminated in a more bizarre fashion. The secret to good health, according to Fletcher, was mastication. Eating too much food without thorough chewing resulted in indigestion and poor assimilation of the food. But if food were chewed until it liquefied in the mouth and automatically slid down the gullet, the ravages of disease could be avoided. So Fletcher chewed and chewed, once taking 722 chews to liquefy a shallot. He claimed that through this method people would be satisfied with far less food, requiring only twelve to fifteen mouthfuls to be satiated. Fletcher himself found that he could easily get by on about 1,600 calories, far less than the average consumption at the time. Particularly noteworthy was the fact that this represented a considerably smaller protein intake than doctors were recommending.

Although Fletcher was at first regarded as a crank, people started to pay attention when this nutritional apostle began to demonstrate amazing feats of strength in spite of his low calorie intake. He beat younger men in bicycle races and routinely bested the strongest university students in weight-lifting contests. So America began to Fletcherize and scientists studied the results. Professor Russell Chittenden of Yale University concluded that Fletcherizers took in far less protein than what at the time was believed to be the optimal amount, and yet improved their strength. He himself claimed that his own headaches and rheumatic knee pain were resolved by following the chewing regimen.

Fletcher's popularity grew and grew and more and more people started to chew and chew. "The Great Masticator," as he was called, spoke to medical societies, had his books translated into many languages, and had his theories discussed in the pages of the foremost medical journals, especially his claim that there would be no slums, no degeneracy, no criminals, and no need for doctors if only the world chewed properly.

And for a while the world tried. Urged by the likes of Thomas Edison, John D. Rockefeller, and Kellogg, people exercised their mandibles like never before. They switched to fruits and vegetables from meat because the latter was hard to liquefy. They ate less and felt better. Slenderness began to be chic. But soon they tired of chewing, and memories of the man who had captivated a whole generation began to fade.

Fletcher perhaps bit off a bit more than he could chew, but his ideas were not completely nonsensical. Since saliva contains enzymes that begin the process of breaking down food components, proper chewing does aid digestion. Food that is properly digested is more readily absorbed, so that in theory we could get away by eating less. People with inflammatory bowel disease may also experience less irritation with thoroughly chewed food. And there is yet another benefit, especially to those afflicted by the excessive passing of "wind." Extensive mastication results in less swallowed air and consequently fewer emissions.

Although Fletcher's emphasis on mastication cannot be wholeheartedly supported by modern science, we must admit that he was one of the first to rail against dietary excess and he was the first to demonstrate that the human requirement for daily protein is actually quite modest. As far as his theory that proper chewing may prevent criminal behavior goes, well, that's a bit hard to swallow. But since we haven't been too successful in this area, perhaps a little Fletcherizing wouldn't hurt. We might even lose some weight in the process. At least according to a study at the prestigious Mayo Clinic in Minnesota.

This engaging bit of research was prompted by the knowledge that cows expend a fair amount of energy by chewing. About 20 percent of the calories "burned" by cows can be ascribed to their ruminations. How do we know this? By comparing the energy expenditure of cows that are fed intravenously to that of cows fed normally. In other words, cows that don't have to spend energy chewing their cud are more likely to put on weight. The researchers wondered if there was a human parallel; that is, could humans possibly lose weight by chewing? Seven adults were fitted with masks that allowed inhaled oxygen and exhaled carbon dioxide to be measured, since calorie expenditure can be calculated from the ratio of these two gases.

The volunteers were asked to sit in a temperature-controlled, dark, silent lab, with their arms and legs supported so that they would not have to exert any effort other than that required by the experiment. Energy expenditure was measured at rest for thirty minutes, then the subjects were given a sample of calorie-free gum and, as reported in the *New England Journal of Medicine*, were instructed to "chew at a frequency of 100 Hz, a value that approximates chewing frequency at our institution." A metronome was used to make sure that everyone was chewing at a constant rate. Just like the cows, there was an increase of 19 percent in energy expenditure while chewing. What does this mean? The scientists concluded that chewing gum during waking hours would lead to a weight loss of five kilograms in one year! But according to their data, we would have to be chewing at a pretty rapid pace to achieve this weight loss. A hundred hertz, in everyday language, means a hundred times a second! Now, either the employees at the Mayo Clinic are an especially talented bunch of rapid chewers or the researchers forgot their elementary physics. I suspect the latter. Undoubtedly they meant a chewing frequency of a hundred times a minute, not a second.

An interesting bit of research to be sure, but I would still recommend other forms of exercise for weight loss. If you are not into exercise though, you may want to take a look at a study carried out at Johns Hopkins University. Here researchers examined how various aspects of people's environment affect their eating habits. They discovered that when walls were light blue or green, 40 percent of people chewed their food more and ate less. On the other hand, when the walls were red, yellow, or orange, chewing went down and food consumption went up. Maybe that's why fast food restaurant designers choose these bright colors! Horace Fletcher would not approve. He would probably chew them out.

INSERT YOGURT WHERE?

It must have been quite a scene. The little man, no more than five foot four, dressed completely in white, center stage, playing pitch and catch with a chimp. But there was no ball in sight. Dr. John Harvey Kellogg was tossing pieces of steak at the monkey, who threw them right back. Then the good doctor repeated the comic event, using a banana. This time his costar didn't return the toss. To the applause of the throng who had filled the great hall at the Sanitarium in Battle Creek, Michigan, the chimp happily ate the banana. "Even a dumb animal knows what it should eat and what not," bellowed the doctor!

Those who were unconvinced that this spectacle proved the benefits of a vegetarian diet were then urged to come up onstage for a more dramatic demonstration. They were invited to gaze through a microscope at a piece of steak and a sample of manure. To their horror, the meat harbored more bacteria than the excrement! After that shocking experience, few complained about the spartan vegetable- and grain-based diet that was the standard fare at "the San."

In the late 1800s, the Battle Creek Sanitarium was unquestionably the place to be for people who needed to be cured of diseases they never had. Kellogg and his staff catered to rich hypochondriacs who were usually diagnosed to be suffering from "autointoxication." Dr. Kellogg was convinced that virtually all illnesses originated in the bowels and that the "putrefaction changes which recur in the undigested residues of flesh foods" were to be blamed for disease. To cure autointoxication, the bowels had to be cleansed. For this Kellogg had at his disposal a variety of enema machines designed to flush the colon with impressive amounts of water in just a few seconds. He often boasted that he himself started every day with an enema! After water had flushed out the nether regions came the yogurt treatment. For both ends. Dr. Kellogg believed that the bacteria used to make yogurt were protective against disease and "should be planted where they are most needed and may render the most effective service."

There were other options for those who did not see the appeal of being pumped full of yogurt through their rear portals. The San's "mechanotherapy" department had come up with the vibratory chair. This was a spring-loaded device that shook the patient violently to stimulate intestinal peristalsis. Once the toxins had been dislodged in this fashion, headaches and backaches would disappear and, according to Kellogg, the body "would be filled with a healthy dose of oxygen." And the San's coffers would be filled with a healthy dose of money.

The San also offered a variety of baths: cold, hot, and electrified. If this did not shock the disease out of the unfortunate victim, then Dr. Kellogg resorted to surgery, removing the offending part of the intestine. Kellogg carried out over 22,000 such operations during his career with a remarkably low complication rate. He was actually a gifted surgeon who had trained at the Bellevue Medical College in New York, where his education was financed by Ellen White, the leader of the Seventh-Day Adventist movement. White had opened a Health Reform Institute based on hydrotherapy and vegetarianism but wanted the place to have medical legitimacy. Kellogg came from an Adventist family and seemed like an ideal candidate to run the institute.

At the age of twenty-four, John Harvey took up the challenge and quickly coined the name Sanitarium for the establishment where he would practice his particular blend of medicine and malarkey for sixty-two years. It was a grand place. Thomas Edison, Henry Ford, S.S. Kresge, and even President William Taft were visitors. They came to exercise in special athletic diapers to the beat of the "Battle Creek Sanitarium March," played by a brass band. They came to be dunked in electrified pools and to have various parts of their anatomy assaulted with streams of water. And they came to be told to eat what the monkey eats — simple food and not too much of it. The doctor maintained that eating meat was sexually inflammatory and would lead to "self-abuse," which robbed the body of vigor and health. Even regular sexual activity was to be curtailed! Kellogg lived by his theories and often proclaimed that he was living proof that sex was not necessary for good health. He had never consummated his marriage! History does not record Mrs. Kellogg's views on this matter.

John Harvey, together with his brother Will Keith, developed a number of foods to replace meat in the diet. He came up with various nut butters, was an early proponent of soy, and looked for various ways in which whole grains could be easily incorporated into meals. The doctor was particularly fond of zwieback, a twice-baked biscuit that he claimed helped the bowels eliminate toxins. As the likely apocryphal story goes, one day an elderly patient broke her false teeth on the hard biscuit and demanded compensation from Kellogg. It was then that the brothers cooked up some wheat and poured the mush between rotating rollers to produce wheat flakes. Cornflakes followed soon after. John Harvey was only interested in the health properties of the new products. Could these serve as an antidote to the passions stirred up by meat? But Will was a businessman and was bent on commercialization. He eventually gained control of the Kellogg company after a feud with John Harvey and turned breakfast cereals into one of the world's greatest success stories.

Dr. John Harvey Kellogg may have been eccentric, but in some ways he was ahead of his time. Menus at the San listed nutritional composition and calorie counts. He insisted that his patients get plenty of exercise and fresh air. Many of his ideas about vegetarianism have been corroborated by modern science and research has shown the potential benefits of consuming foods containing certain types of live bacteria. Indeed I've started to supplement my breakfast of cornflakes, flaxseeds, and blueberries with some live culture yogurt. Only via the oral route.

DINSHAH AND THE SPECTRO-CHROME

When I was young and had a sore throat, my mother would always wrap a scarf around my neck. It felt warm and comforting, but I don't think it did much therapeutically. That may be because it was the wrong color. At least, that is what the manufacturers of the Healing Scarf would undoubtedly suggest. This medical miracle caught my eye on the Internet because of its claim to "balance your energies and increase your sense of well-being." It is a pretty scarf, to be sure. Made of Chinese silk, it features all the colors of the rainbow and is "designed to bring all healing colours into your consciousness."

Healing with color? Where did they get this idea? I decided to track it down. I'm glad I did, for the winding path led me to one of the most fascinating characters in the history of scientific quackery. Let me tell you about Dinshah P. Ghadiali and his Spectro-Chrome. Dinshah, as he liked to be called, was born in India in 1873. By his own account, he was a remarkable man. He began school at the ripe age of two and a half, by eight was in high school, and by eleven he was an assistant to a professor of mathematics at a college in Mumbai. Dinshah recounts that he began to study medicine at the age of fourteen, but then we hear no more about his progress in this area. Probably because he saw no need to pursue these futile studies once he had independently discovered the key to health. Color therapy.

Dinshah apparently came upon this discovery when he cured a young girl dying of colitis by exposing her to light from a lamp fitted with an indigo-colored glass filter. The therapy also involved giving the patient milk that had been placed in a bottle of the same color and exposed to sunlight. Within three days, the girl was well and a career was launched. Dinshah opened an Electro-Medical Hall in India, where he began to refine his treatment. By the time he came to America in 1911, he had a theory, albeit a bizarre one, to go along with his colored lights. Every element, he said, exhibits a preponderance of one of the seven prismatic colors. Oxygen, hydrogen, nitrogen, and carbon, the elements that make up 97 percent of the body, are associated with blue, red, green, and yellow. In health these colors are balanced, but in disease fall out of balance. Therapy is simple; to cure disease administer the lacking colors or reduce colors that have become too brilliant.

Of course, Dinshah had exactly the method to use. His Spectro-Chrome was a box with a light bulb and an opening that could be fitted with various colored filters. It was accompanied by the Spectro-Chrome Therapeutic System guide, detailing the appropriate colors to shine on a patient. Green light, for example, was a pituitary stimulant and germicide, while scarlet was a genital stimulant. Any disease, save broken bones, was amenable to color therapy. The Spectro-Chrome was especially suited for use by intelligent people, Dinshah said, because "drugs quickly upset the nervo-vital balance of persons of high mental and spiritual development." A pretty clever trap. The gullible, thinking themselves to be intelligent, ate it up.

To many people the argument about the benefit of color therapy seemed convincing. After all, they knew that premature babies were treated with blue light to cure them of jaundice, that sunlight was needed for the synthesis of vitamin D in the body, and that plants absolutely required light for growth. Add to this the notion that chemists had shown that elements when heated emitted different colors of light, and Dinshah's preposterous notions seemed to make sense. His slogan of "No Diagnosis, No Drugs, No Surgery" also sat well with a public largely unsatisfied with current medical care. The non-invasive therapy and the promise of a cure for virtually any ailment was very appealing.

Of course it wasn't long before Dinshah ran into trouble with the establishment. He was labeled a fraud and a charlatan by the American Medical Association but managed to cunningly portray himself as a humanitarian who was being persecuted by the money-grabbing, ineffective, jealous physicians. To protect himself legally, Dinshah came up with some incredible lingo. He didn't talk of cures, he spoke of "normalating" the body. Instead of treating patients he claimed "to restore their Radio-Active and Radio-Emanative Equilibrium." This would be done with his light exposures, or "tonations." Tonations would be carried out with the patient lying with his head to the north, so as to align the Earth's and the body's magnetic fields, of course. Dinshah also designed Spirometer Rods to measure the pressure difference between the two nostrils to determine when during the day tonations should take place to take advantage of the body's natural tides. Special thermometers applied to the bare skin above the organs would determine if a condition was acute or chronic and what kind of light therapy was needed. It would be hard to think of a more convoluted and irrational form of therapy.

In 1931 Dinshah had his first run-in with the law over the Spectro-Chrome. Actually he had a skirmish six years before when he was arrested for transporting a nineteen-year-old girl, his secretary, across state lines for immoral purposes. Perhaps he had been exposing himself to too much scarlet light. In any case, he spent four years in jail. But now he was arraigned on second-degree grand larceny after being charged by a former student who claimed that the Spectro-Chrome did not perform as promised. Dinshah trotted out numerous satisfied patients in his defense, incredibly including some physicians. In fact, a surgeon, Kate Baldwin, who claimed that she had successfully treated glaucoma, tuberculosis, cancer, syphilis, and a very serious burn case. The government had experts testifying that the Spectro-Chrome was nothing other than an ordinary lamp and that the successes were all due to the placebo effect. The prosecution could not prove the intent to defraud, and Dinshah was found not guilty. He went back to selling more Spectro-Chromes, now claiming that he had been vindicated by the trial's outcome.

After the passage of the Food and Drug Act of 1938, which gave the FDA some teeth in regulating therapeutic devices, the government began to assemble evidence against Dinshah. Finally in 1945 he was charged with introducing a misbranded article into interstate commerce, a violation of the criminal code. Once again he trotted out his satisfied patients, but this time there were no supporting physicians. His fate was virtually sealed when one of his star witnesses, whom Dinshah had "cured" of seizures, had a fit on the witness stand. The jury also heard how patients he claimed had been cured had actually died. The prosecution brought a witness who Dinshah had repeatedly profiled in his advertising as having been cured of paralysis. She could not take a single step when the master urged her. And finally the court heard how the celebrated burn victim, described as a miracle cure by Dr. Baldwin in the previous trial, had in fact died of her injuries. Another witness described how he had called Dinshah after his diabetic father had lapsed into a coma and was told to just shine the yellow light on him. He did, until the man died.

Dinshah was heavily fined, his books and lamps were seized, and he was put on five years' probation. But he was persistent. The day after his probation ended, he was at it again. This time he founded the Visible Spectrum Research Institute and sold lamps labeled as having "no curative or therapeutic value." Of course he implied in his literature that this was only to keep the FDA dogs away — just meaningless legalese that David had to resort to against Goliath. In

1958 the government obtained a permanent injunction against shipping Specro-Chromes across state lines, but the persistent Dinshah kept selling them in New Jersey. After his death in 1966, his sons took over the reins and managed to have the Dinshah Health Society of Malaga registered as a non-profit, scientific, and educational organization that is tax exempt. This non-profit organization sells all kinds of light therapy books, including a history of the Specro-Chrome by Dinshah himself in seven clothbound volumes for $220. You can of course also buy plans to build an inexpensive Spectro-Chrome from a light bulb, cardboard, and colored gels. They apparently do not sell the finished product, but another company on the Internet does advertise Color Light Therapy Lamps "as recommended by Dinshah." These look suspiciously like theater spotlights with colored gels.

It seems that today there is still enough ignorance about what light is, about disease processes, and about how the body functions to allow the gullible to be victimized. The colorful Dinshah may have lived in what we think are enlightened times, but his pseudoscientific ideas smacked of the Dark Ages. I don't know if he would have approved of the Healing Scarf, but I suspect it would have been right up his alley. The scarf though, I must admit, is so pretty that I bought one. It works. It keeps my neck warm.

ELECTROQUACKERY

The late 1800s introduced miracle after miracle to the American public. Marconi's radio, Bell's telephone, and Edison's light bulb ushered in the age of electricity. If this mysterious force could send sound through the air and dispel darkness, might it not also work its magic in the human body? Scientists began to address this question in earnest, but long before they could cast light upon the situation, the quacks and charlatans entered the game. Unencumbered with the need to play by the rules of science, their unsubstantiated claims and pseudoscientific lingo sparked plenty of interest.

Galvanic Electric Belts claimed to cure "nervous and chronic diseases without medicine." Filled with primitive batteries consisting of pieces of copper and zinc separated by blotting paper, they delivered a mild current that convinced the gullible that the healing process was indeed underway. One of the most popular belt designs featured loops extending down to the testicles. This was aimed at restoring "lost manhood," a loss the manufacturer attributed to "the greatest outrage on Nature's sexual ordinances man can possibly perpetrate," namely self-pleasure. Perpetrators of this transgression against nature could be identified by black-and-blue discolorations under the eyes. But luckily, they could be reenergized and dissuaded from such activities by wearing a Galvanic Belt.

And for those who were leery of electrical equipment, there was Electric Liniment or pills which "contained 50,000 volts of electricity in a 2 drachm bottle." The only charge patients got out of these nonsensical nostrums appeared on the quack's bill!

Surely the dean of the electrical quacks was Dr. Albert Abrams, a traditionally trained physician who practiced standard medicine, and for a while in the early 1900s served as vice president of the California Medical Society. As he approached middle age, Dr. Abrams decided that standard medicine did not really suit him. So, in 1909 he invented his own specialty, which he called "spondylotherapy." There was no longer any need to rely on symptoms or stethoscopes for diagnosis, because Abrams had decided that he could tell what the problem was by how a patient's spine resonated when tapped. After the diagnosis had been made, the cure quickly followed with an appropriate percussion of the spine.

The widespread introduction of electricity was tailor-made for Dr. Abrams. Now he could put his vibrational ideas on a scientific footing! Diseases, he suggested, were caused by a disharmony of electronic oscillations in the body and could be cured by vibrations that have the same frequency as the disease. He invented a device, known as the dynamizer, to diagnose illness by measuring electronic vibrations in a drop of blood. The diagnosis did not even require the presence of the patient, but did require a healthy surrogate. Just picture the bizarre scene.

A few drops of blood from the patient were treated with a large magnet to "cleanse" them of confusing vibrations and then were introduced into the dynamizer. A wire from this machine was connected to the forehead of a healthy volunteer who stood on a metal plate. Abrams proceeded to systematically tap the surrogate's body until he located an area that was somehow in resonance with the vibrational frequencies of the blood sample. Thus the diseased organ was identified and, incidentally, also the patient's religion. Abrams then used a second machine, called an oscilloclast, which he tuned to the vibrational frequency of the disease to cure it. Testimonials sang the praises of Abrams's brilliant machines and the money began to flow in.

The noted physicist Robert Millikan called the oscilloclast a device a ten-year-old boy would build to fool an eight-year-old. The *New York Times* called spondylotherapy a scheme of magnificent absurdity. The American Medical Association produced posters claiming that Abrams's disciples diagnosed nonexistent diseases and then made a fortune by treating them.

But the good doctor was undeterred. Business flourished. When prohibition came along, he introduced a gadget that could duplicate the vibrational frequency of alcohol so Abramsites could get drunk without drinking. More testimonials followed. Until someone had the idea of investigating the absurd claims. It was easy enough to do. The public began to be skeptical when Abrams diagnosed general cancer and tuberculosis of the urogenital tract in a sample of chicken blood. And interest really waned when Abrams himself contracted pneumonia and died from the disease his oscilloclast was supposed to cure easily.

Of course quack electrical devices did not die with him. In fact, through the electronic marvels of the Internet, they have proliferated.

You can get yourself a Medicomat, which will treat asthma, arthritis, and hepatitis, an Interro device that will diagnose "imbalances" in the body, or a Q-LINK Pendant to guard against "toxic forms of energy," which consists of a plastic case, a coil of copper wire, and a computer chip. A bargain for $129 U.S. Then there is the Crystaldyne pain reliever, guaranteed to work for pain ranging from arthritis to menstrual cramps. Well, I just had to order one of those. What I got for my $50 of "research funds" was a $2 barbecue grill igniter. Eliminating pain was to be as simple as pressing against the skin and pushing the button. It came in handy; I used it to replace the nonfunctioning igniter on my barbecue.

THE PRINCE OF HUMBUG

There is no evidence that he ever uttered the quote for which he became famous. Indeed, it is very unlikely that Phineas Taylor Barnum would ever have stated that "there is a sucker born every minute." The "Greatest Showman" did not look on the throngs that flocked to Barnum's American Museum, eager to see his exhibits of curiosities, as suckers. The folks may have been drawn by some "humbug," but Barnum insisted that they got their money's worth of entertainment.

Visitors may have been attracted by the giant picture of the "FeeJee Mermaid" that adorned the front of the museum, but once inside were confronted with the skull of a monkey cleverly attached to the skeleton of a fish. However, even those who recognized this as a hoax never complained because they were treated to exhibits of exotic live animals, a flea circus, magic acts, glassblower demonstrations, ventriloquists, and performances by Grizzly Adams's trained bears. They could have their skulls examined by phrenologists or meet General Tom Thumb, who stood all of 25 inches tall and had charmed Queen Victoria. Chang and Eng the Siamese twins were there, along with Josephine Clofullia, the bearded lady who had a genuine case of hirsutism.

Barnum did not mind that some of his exhibits aroused skepticism; in fact he welcomed it. Visitors' doubts fostered debates and resulted in more publicity and increased ticket sales. When the bearded lady didn't generate the interest he had hoped for, Barnum hired a man to sue him for false advertising, claiming Josephine was really a man dressed up as a woman. The case went to trial, in which Barnum introduced three physicians who had examined Josephine and testified that she was indeed a woman. The press reported on the dismissal of the lawsuit and visitors surged through the turnstiles to meet the "authenticated" bearded lady. None thought of themselves as suckers, except perhaps those who had been taken in by signs indicating "this way to the Egress" and found themselves outside. Barnum was concerned people were lingering too long at the exhibits and took advantage of some not recognizing that "egress" was another term for "exit."

The "Prince of Humbug's" rise to fame had an unsavory path. It all began when at the age of twenty-five he got wind of a traveling act in Philadelphia that featured a woman who supposedly had been George Washington's nursemaid and was now 161 years old! Barnum purchased the act and shamelessly exhibited the toothless, blind, semi-paralyzed, formerly enslaved woman, African American Joice Heth, as the "greatest natural and national curiosity in the world." In response to doubts about her age, Barnum announced that when she died, Heth would be autopsied to prove her longevity. He followed through, hired a surgeon and charged people to watch the gruesome performance. The doctor declared that the age claim was a fraud but Barnum got the publicity he wanted.

"Zip the Pinhead" with his tapering cranium was another one of Barnum's sideshow displays as the "missing link," supposedly discovered in Africa. William Henry Johnson, who was well paid, was a big attraction in the museum, where dressed in a furry suit he would rattle the bars in his cage and screech at passersby. Later in his career,

Barnum entered the political arena, expressed regret over his abuse of Heth and Johnson, and became a vocal opponent of slavery.

While Barnum's sordid history as a showman has been addressed by historians, far less attention has been paid to his exposés of medical quacks and various other charlatans. His 1866 book, *The Humbugs of the World*, is a classic in which he skewers hoaxers of all sorts and exposes the antics of fake mediums some fifty years before Houdini achieved fame with his revelations about séance scams. Indeed, Barnum was the first to offer a reward to any medium who could prove an ability to communicate with the dead. He never had to come up with the $500 prize.

Fraudsters such as the Davenport brothers and the Fox sisters, who claimed to contact the spirit world, also came under scrutiny. The Davenports allowed themselves to be tied up inside a "spirit cabinet" along with some musical instruments. When the cabinet's doors were closed, the instruments began to play, supposedly by spirits. Barnum explained to his readers the tricks that the brothers used to free themselves of the bonds to play the instruments themselves. The "rappings" heard at the Fox sisters' performances that were said to be communications from the spirit world were nothing other than the girls cracking their toes.

Barnum also condemned merchants who adulterated food, a common practice at the time. He revealed how milk often had a component that came from a "cow with a wooden tail," a reference to the handle of a water pump. Coffee was adulterated with chicory root and dark tea was converted to the more prized green version with Prussian blue. He advised people to stay away from processed foods, long before this became fashionable, saying that you "can't adulterate an egg, nor an apple or a potato."

The most vicious attacks were reserved for the medical quacks, such as a Dr. Andrews whose Hasheesh candy was a "sure cure for all diseases of the liver, stomach and brain." While dismissing these as odious deceptions, Barnum also invoked the idea of the placebo effect, actually using that term.

While P.T. Barnum's reputation as a scoundrel who swindled people with his hoaxes has some merit, he curiously was also a champion of critical thinking. He railed against "blatherskites," people who talk but make no sense, and proposed that "it is high time that the credulous portion of our community should be saved from the deceptions, delusions and swindles of blasphemous mountebanks." We need to conjure up his spirit.

MCFADDEN AND PHYSICAL CULTURE

Bernard McFadden had no use for doctors. They couldn't cure any disease, he maintained, but could certainly cause plenty of misery. And he wasn't completely wrong. Back in 1884, when sixteen-year-old Bernard sought treatment for his hacking cough and digestive disorders, the treatments offered were pretty abysmal. Bloodletting was still practiced, in spite of the fact that it tended to convert patients into corpses. Purging with mercurous chloride was hardly better, often resulting in a bloody evacuation of the bowels, to say nothing of bleeding gums and mouth sores. By contrast, tobacco smoke enemas and goose grease poultices must have seemed positively pleasant. If you got better, it was often in spite of the treatments, not because of them. Little wonder then that one frustrated patient referred to physicians as "God's nutcrackers," seeing that "they opened the corporeal shell to let the soul escape."

McFadden's frustration with medicine caused him to take matters into his own hands. Exercise, Bernard concluded, was the key to health, and he began to spread the word about "the wild joy that thrilled my nerves when I began to feel that health and strength were surely within my reach." Believing that people would pay more attention to his advice if he spiced up his name, Bernard McFadden became Bernarr Macfadden, guru of "physical culture," and author of the slogan "weakness is a crime, don't be a criminal." Macfadden's enthusiasm for exercise and his anti-doctor rants struck a chord with a public hungry for simple solutions to complex problems. His *Physical Culture* magazine was eagerly gobbled up, and by the 1930s Macfadden had built a multimillion-dollar publishing empire.

The exercise maven sure knew how to garner publicity. He sponsored a contest to find the most physically perfect woman in England and then proceeded to marry her. Little did Mary Williamson realize what she was getting herself into. Jumping onto Bernarr's stomach from a high table to demonstrate his fitness in public was just the beginning. She was also expected to buy into some of the more bizarre theories her husband had begun to espouse. "Drugs never cured anything unless you call death a cure," he proclaimed. Diseases were caused by "impurity of the blood" and the cure was fasting. The body, deprived of nutrients, was forced to devour the "impurities." Mary had to endure her children being fasted for whooping cough, measles, and chicken pox, and to suffer the ultimate tragedy of watching a son die as Bernarr attempted to cure him of convulsions by plunging the boy into hot water. Other strange ideas followed. Forget the dentist, Macfadden advised, just chew on wood for aching teeth. "Air baths," which amounted to walking around the house naked, were great for health. The male genitals were to be exercised with an instrument of his invention, called the peniscope.

To Macfadden's credit, he decried the use of tobacco and alcohol, taught the superiority of whole grain products, and even started a chain of physical culture restaurants where vegetarian soups, beans, chopped nuts, and vegetable steaks were served. Unfortunately, Macfadden maintained his anti-doctor diatribes into the 1950s, even as medicine began to make significant advances. Still, there is no doubt that his urging the public to engage in exercise reaped benefits, benefits which are being confirmed by modern science.

Our immune system, research shows, can be revved up by exercise. People who engage in physical activity catch fewer colds. A study at the University of South Carolina found that among 550 healthy men and women, those who got at least moderate exercise most days averaged one cold a year, while people who were less active reported four colds annually. Even recovery from colds may be hastened in people who follow a regular exercise program. David Nieman of Appalachian State University studied women with colds and found that ladies who walked regularly and caught colds recovered in less than five days, while those who were sedentary suffered for seven days. So, it seems Macfadden was really onto something with his claim that he never caught a cold because he walked twenty miles to work every day. He must have walked briskly though, given his theory that one should wear as little clothing as possible, even in winter. Indeed, Bernarr favored going commando.

Exercise is great, but more isn't necessarily better. Marathon runners have a greater chance of catching colds for several days after a race. Even with heart health, where the benefits of activity have been clearly demonstrated, it seems that exhaustive amounts of exercise are not needed. Jogging or walking quickly for 12 miles a week confers significant cardiovascular benefits. Extending this to 20 miles while maintaining the same intensity provides even more benefits. But there is no need to constantly strive for increased speed.

Exercise also has a role in the fight against that most dreaded of all diseases, cancer. Women with breast cancer who exercise at moderate levels for three to five hours a week have been shown to have higher survival rates than women who exercise less than an hour a week. Studies have also shown that regular physical exercise can slow the progression of prostate cancer. And if you are still not convinced of the value of exercise, consider that it may help keep Alzheimer's disease and other dementias at bay. Researchers at the Karolinska Institute in Sweden checked for the incidence of dementia or Alzheimer's disease in seniors whose exercise habits were monitored for over thirty years. Those who engaged in moderate physical activity as they passed through middle age had a 50 percent lower chance of developing dementia and a 60 percent smaller risk of Alzheimer's than their sedentary confreres. Macfadden himself was a shining example of the merits of exercise. He earned his pilot's license at age sixty-three and celebrated his eighty-fourth birthday with a parachute jump. At the age of eighty he married a forty-three-year-old. History does not record whether he needed to make use of his peniscope.

GETTING STEINACHED

The Great War was over, cars were multiplying on the streets, radios were crackling in living rooms, plastics were hitting the market, and theaters were attracting people with newfangled moving pictures. Science and technology were roaring ahead. It was, after all, the "Roaring Twenties." But in Vienna, there was another kind of roar. It was emanating from thousands of older men who claimed to have regained their virility through what seemed to be a stunning advance in medicine. They had been "Steinached"! The men had undergone a twenty-minute procedure introduced by Dr. Eugen Steinach in which one of their seminal ducts was tied off. In other words, the men underwent a partial vasectomy. The goal wasn't prevention of pregnancies, it was rejuvenation!

Steinach's work was inspired by French physiologist Charles-Édouard Brown-Séquard's seminal lecture, delivered to members of la Societé de biologie in 1899, in which he described having injected himself with filtered extracts from the crushed testicles of young dogs and guinea pigs to regain the vigor and intellectual stamina of his youth. The professor had also tested himself with a dynamometer, a device that measures mechanical force, and found that his muscle strength had been renewed. He capped off the lecture by telling his rapt audience that just hours earlier he had passed the final test of his experiment by "paying a visit" to his young wife.

The scientific community, however, did not buy Brown-Séquard's claim that the key to rejuvenation was injection of minced gonads. The prestigious *Boston Medical and Surgical Journal* opined that "the sooner the general public and especially septuagenarian readers of the latest sensation understand that for the physically used up and worn out there is no secret of rejuvenation, no elixir of youth, the better." Biology professor Steinach, however, thought that Brown-Séquard's work with gonads was worth pursuing and turned to transplanting the testes of a male guinea pig into a female. She then exhibited mounting behavior characteristic of a male! Steinach concluded that the gland's secretions were responsible for sexuality and even theorized that homosexuality in men could be treated by transplanting a testicle from a "normal" man into a recipient in need of "remasculinization." Thankfully that idea didn't fly, but surgeon Serge Voronoff's notion of grafting monkey gland tissue onto the testicles of aging men did!

While serving as physician to the king of Egypt, Voronoff had noted that the court eunuchs were often sickly and seemed to age very quickly. The testes, he concluded, played an important role in maintaining vigor, and that "possession of active genital glands was the best possible assurance for a long life." In 1918 he believed he made his point when he restored an aging ram's youthful vitality by transplanting the testes of a young lamb.

Voronoff upped the ante by transplanting the testes of executed criminals into aging men rich enough to pay for the procedure. But demand soon outstripped supply, and since few young men were willing to part with their precious parts even for rich compensation, Voronoff came up with an alternate scheme. He would transplant bits of chimpanzee and monkey testes onto the genitals of elderly men. Eventually more than a thousand men underwent the monkey gland treatment at the hands of doctors around the world, with the requisite material often being supplied by a monkey farm Voronoff set up on the Italian Riviera.

Steinach bought into Voronoff's idea, but thought that the benefits ascribed to transplants could be achieved by an alternate procedure. Damming the seminal canal would stimulate the testes to produce more male hormones! At the time, researchers had determined that there were two types of tissues in testicles. Seminal tubules produced spermatozoa, but there were are also Leydig cells between the tubules that released sex hormones. Steinach's idea was that the two types of tissues compete for nourishment and that stifling the sperm-producing tissues would boost the production of the sex hormones.

In his book, *Sex and Life*, Steinach described how his patients "changed from feeble, parched, dribbling drones, to men of vigorous bloom who threw away their glasses, shaved twice a day, dragged loads up to 220 pounds, and even indulged in such youthful follies as buying land in Florida." He believed in his procedure so strongly that he "thrice reactivated himself." It isn't clear what he meant by "thrice," because once the duct is tied off, it's tied off. Whatever improvement Steinach and his patients felt was likely due to wishful thinking because, as we now know, vasectomies do not boost hormonal output by the testes.

Steinach had testimonials galore, including from some very famous people such as Sigmund Freud, who underwent the procedure when he was sixty-seven years old, hoping to improve his "sexuality, his general condition and his capacity for work." William Butler Yeats, the famed writer, was Steinached when he was sixty-nine. "It revived my creative power," wrote Yeats in 1937. Apparently in more than one way. The doctor who performed the snip invited a woman half Yeats's age to dinner with the aim of allowing the writer to make a connection and test out his newly embellished virility. It seems the outcome was successful, with Yeats publicly reporting on his "second puberty" and leading to the Dublin press nicknaming him the "gland old man."

While it is now clear that Brown-Séquard, Voronoff, and Steinach promoted procedures that did not have the claimed efficacy, they did lay the foundations for further research that resulted in the isolation of testosterone, the male sex hormone. In 1927, University of Chicago chemistry professor Fred Koch isolated 20 milligrams of a substance from 20 kilos of bull testes that remasculinized castrated roosters, pigs, and rats. By 1935, German biochemist Adolf Butenandt had worked out the molecular structure of testosterone, the active compound in the bull testes, which allowed him to come up with a chemical synthesis from cholesterol. Today, testosterone and various derivatives are prescribed to men with low blood levels of testosterone, who often claim to experience the effects that were thirsted for by men who subjected their privates to the scalpels wielded by Drs. Steinach and Voronoff in the roaring twenties.

THE DEAN OF QUACKS

The remarkable events about to be chronicled here probably would not have unfolded if young Dr. John Brinkley had not been hired as house doctor at the Swift & Co. meat-packing firm in Kansas in 1917. It was there that he became dazzled by the vigorous mating activities of the goats destined for the slaughterhouse. A couple of years later, when Brinkley had gone into private practice in Milford, Kansas, a farmer by the name of Stittsworth came to see him complaining of a sagging libido. Recalling the goats' frantic antics, the doctor semi-jokingly told him that what he needed was some goat glands! The farmer's response was almost instantaneous: "So, Doc, put 'em in. Transplant 'em."

Most doctors would not have paid any attention to the bizarre request; but then again, Brinkley was not like most doctors. In fact he wasn't a doctor at all. Although he did spend three years at Bennett Medical College in Chicago, he never graduated. He called himself a doctor on the basis of a purchased $500 diploma from the Eclectic Medical University of Kansas City, Missouri, and, as absurd as it sounds, this gave him the right to practice medicine in Arkansas, Kansas, and a few other states as well.

Purchasing a degree from a diploma mill was not out of character for Brinkley. He had worked as a snake oil salesman in a road show and then, together with Chicago con man James Crawford, established Greenville Electro Medical Doctors and made a living by injecting people with colored distilled water for $25 a shot. And that was big money in those days! So Brinkley had all the requirements for capitalizing on the farmer's idea of goat transplants: he was unethical, had a wobbly knowledge of medicine, and he had been captivated by the rambunctious behavior of goats. But Brinkley likely had something else as well. Knowledge of experiments that had been carried out in Europe by Brown-Séquard, Voronoff, and Steinach since the late 1800s.

Brinkley went to work and implanted a bit of goat gonad in Stittsworth's testicle. Within weeks the man was back to thank the doctor for giving him back his libido. And when his wife gave birth to a boy, appropriately named Billy, the farmer began to spread the word of his success. Soon Brinkley's business was booming. Testimonials were pouring in. So was the money. Brinkley was charging $750 per transplant and could not keep up with the demand. All men needed the Brinkley Operation, he maintained, but it was especially suited to the intelligent and least suited to the "stupid type." This of course meant that few men were willing to admit that they had not benefited from the operation.

There were a few problems. Like when Brinkley decided to use Angora goat testicles instead of those from the more common Toggenburg goats. The men who got these were decidedly unhappy. Brinkley himself noted that they reeked like a steamy barn in midsummer. But Brinkley's major problem was that as his fame grew, so did the criticism from the medical community.

Morris Fishbein, the editor of the *Journal of the American Medical Association*, called him a smooth-tongued charlatan and urged that his right to practice be revoked. The claims that the procedure could cure conditions ranging from insanity and acne to influenza and high blood pressure amounted to quackery, Fishbein said. Brinkley wagged his tongue right back and called the American Medical Association a "meat-cutters union." and said that its members were jealous because they were losing business. He then got himself more ammunition by performing a transplant in Los Angeles on Harry Chandler, the owner of the *Los Angeles Times*. The satisfied patient gave Brinkley lots of free publicity.

While in California, Brinkley also learned about the potential of radio. Back home, in 1923, he started up radio station KFKB, whose 1,000 watts were amazing at the time, and featured music, his lectures on rejuvenation, political broadcasts, and his "Medical Question Box," during which he answered listeners' questions. Perhaps the earliest advice show on radio. But the advice dispensed was ridiculous and usually involved giving a prescription by a number. This could be filled at a local pharmacy; Brinkley had set up the National Dr. Brinkley Pharmaceutical Association in collusion with pharmacists who relished making lots of money by selling water colored with indigo.

Kickbacks from this operation and continued transplant surgeries made Brinkley into an immensely wealthy man. For $5,000 he would even implant genuine human glands which he obtained from prisoners on death row. He had mansions, a fleet of Cadillacs, airplanes, and yachts. Of course, what he did not have was scientific respect. The American Medical Association finally prevailed upon the Kansas Board of Medical Registration to remove his license for reasons of immorality and unprofessional conduct, and the Federal Radio Commission shut down KFKB for promoting fraud. Brinkley still did not capitulate. He claimed he was being crucified and kept his hospital going by hiring licensed physicians. He purchased radio station XERA in Mexico and began beaming his message into the U.S. with a power of 100,000 watts!

The "Doctor" also decided the only way to get his license back was to become governor. So he organized a massive write-in campaign in 1930 and almost won! His proclamation that he was being persecuted by well-to-do doctors and politicians appealed to people, as did his promise to build free clinics and cure virtually all disease. But Brinkley couldn't even cure himself. The Milford Messiah, as he had been known in his time, the man who had performed over 16,000 goat testicle transplants, the man who appropriately wore a goatee all his life, developed a blood clot and had his leg amputated. Till the very end, his scheming mind was active. Confined to bed, he decided to study for the ministry and had visions of becoming a big-time preacher. He never made it. When the end came, his last words reportedly were: "If Dr. Fishbone goes to heaven, I want to go the other way." If there is any justice, he did.

COLD SHOWERS AND WARM BATHS

Charles Darwin was one of the greatest thinkers of all time. But thinking did not come easy to the father of the theory of evolution and natural selection. Most of Darwin's adult life was plagued with illness; he was constantly tormented by insomnia, palpitations, vomiting, and flatulence.

The doctors of the day had no relief to offer, and in all probability modern medicine would fare little better. Although posthumous diagnosis is always difficult, Darwin's symptoms smack of an insect-borne infection known as Chagas disease. History does record that the great naturalist's first period of ill health was precipitated by a bug bite during a trip to South America.

Today, we know that the infecting organism can be found in the bloodstream for years after initial exposure, wreaking slow havoc and eventually causing heart disease. Indeed Charles Darwin succumbed to a heart attack in 1882.

So did Darwin calmly accept the doctors' failure to provide relief from his affliction? No. Like many others of the day, he turned to alternative therapies, going in big for the "water cure." Darwin was a frequent visitor to "Dr. Lane's delightful hydropathic establishment at Moor Park," where the delights included being plunged into cold baths and being wrapped in cold wet towels. These "hydropathic" establishments were quite the rage at the time.

People in England flocked to trendy establishments such as Harrogate, where the sulphurous waters supposedly had added benefits. Harrogate, however, fell out of favor with ladies of the upper classes when their skin turned a disturbing black after bathing. It seems the ladies had been using bismuth carbonate to whiten their faces so as to distinguish themselves from the working classes with their ruddy complexions. Hydrogen sulphide in the water combined with the bismuth carbonate to form insoluble black bismuth sulphide, which deposited on the skin.

The water cure fad can be traced back to a young Austrian farmer and a runaway cart. Vincenz Priessnitz broke some ribs when the cart crashed into him and found relief from the terrible pain by wrapping himself tightly in a wet bandage. Neighbors began to talk of his amazing recovery, and within a short time Priessnitz had opened a water cure establishment in his hometown of Gräfenberg. By the 1840s he was host to thousands of patients from around the globe. They came to be cured of ailments ranging from nervousness and asthma to smallpox and syphilis. Soon water cure enterprises peppered the landscape in Europe and North America, offering hope to the hopeless.

Treatments varied. Some patients were just wrapped in cold sheets, others were dumped into freezing water as many as thirteen times a day. Priessnitz favored pouring icy water from a height of six meters onto the heads of patients, who had to hold handrails to steady themselves against the pressure. Then there was the "ascending douche," which blasted a stream of water up from the floor, aimed at the genitals. Brrr . . .

Father Sebastian Kneipp, a Bavarian priest, encouraged everyone to stand in cold brooks as long as they could and to bathe in freezing rivers for optimal health. This actually was not a new idea. Almost a hundred years earlier, the Scottish poet Robert Burns had been encouraged by his physicians to stand for two hours a day up to his armpits in the freezing waters of the Solway Firth. He died at the age of thirty-seven.

As can be expected for any treatment, no matter how outrageous, hydropathy resulted in numerous glowing testimonials. Modern science would ascribe the perceived benefits to the placebo effect or to the spontaneous resolution of the ailment. But by and large the water cure could not deliver on its promise, and by the end of the 1800s, the movement had essentially petered out, destined to be relegated to a footnote in the annals of medical history.

The present rekindling of the infatuation with "alternative therapies" has, however, resulted in a reincarnation of water therapy. It seems that there is never a dull moment in our search for novel ways to battle adversity. Enter Harold Dull, the inventor of Watsu. This gentleman had learned all about Zen shiatsu, a technique that combines the application of pressure to distinct points on the body using the fingers and the palms with various muscle stretches. The idea is to improve health by unblocking the channels through which qi flows. Qi is the Chinese term for some sort of life force that surges through the body. Since it is unmeasurable by any method known to science, it is indeed difficult to know if it flows more freely after a Watsu treatment.

Dull somehow concluded that the technique would be even more effective if done under water. When the maneuvers are performed in this fashion, "deep emotions which keep us from fully experiencing life and which rob us of the natural power of life's energy surface . . . are washed away by the surrounding water." The treatment is also said to build trust in humanity, as the therapist and client communicate through the rhythm of their bodies. It must at least build trust in the therapist, who has to keep the client's head out of the water with one hand while swinging the knees open and closed with the other.

Since the therapist is also in the water, it is understandable that Dull wanted nothing to do with cold water. Watsu (the term is a contraction of water and shiatsu) is performed in water with a temperature of 34°C! All the new-age mumbo jumbo aside, I'm sure that Watsu is a pleasant and relaxing experience. A lot better than being doused with cold water.

But wait a minute! What did they recently discover at the Thrombosis Research Institute in London? Could this be? Volunteers who took cold showers were better able to fight off viral infections such as the cold or the flu! Not only that, but the subjects reported harder nails and improved hair growth. There was even some evidence for a beneficial effect in chronic fatigue syndrome.

And perhaps of greatest interest was the observation that the cold temperatures produced an increase in hormones that regulate potency in men. So perhaps those nineteenth-century hydropathists weren't so crazy after all. Maybe we should think about resurrecting some of those great water cure establishments! But to tell you the truth, warm Watsu sounds more inviting than a cold shower. I think I'll go think it all over in a lukewarm bath.

A RABBIT OUT OF A HAT

Pulling a rabbit out of a hat, well that's sort of old hat. But what about pulling a rabbit out of a woman's privates? That is exactly what happened in 1726 when Mary Toft, an illiterate servant, seemed to give birth to a litter of rabbits and assorted other animal parts. Incredibly some physicians were taken in by the hoax, which captured the imagination of England as well as that of King George I.

The remarkable story began when Dr. John Howard was called to Mary's home to assist in her labor, but instead of a baby, she delivered what looked like animal parts. That didn't end the pregnancy though, and Howard was repeatedly called back over the next month to deliver first a rabbit's head, then the legs of a cat, and finally a litter of nine dead baby rabbits. Howard didn't know what to make of this bizarre phenomenon and sought help from other doctors. When the king heard about the story, he immediately dispatched surgeon Nathaniel St. André to look into the matter. Much to his surprise, St. André also witnessed Mary giving birth to several dead rabbits and became convinced that some sort of supernatural event was occurring. King George thought that further investigation was warranted and sent German physician Cyriacus Ahlers to investigate. He too witnessed several rabbit births but smelled a rat. One of the rabbits still had dung pellets inside that contained corn and hay. Since Mary's uterus was unlikely to produce such crops, Ahlers reported to the king that he suspected a hoax.

By this time the story had become a national sensation, prompted by St. André's publication of *A Short Narrative of an Extraordinary Delivery of Rabbets* and Mary's curious explanation for her strange prodigies. She claimed to have been startled by a rabbit while working in a field, which caused her to have a strong desire for rabbit meat, but being very poor, she was unable to satisfy the cravings. People bought the story because at the time there was a belief that emotional disturbances during pregnancy could influence the developing fetus.

"Maternal impression" still held sway into the nineteenth century, exemplified by the case of Joseph Merrick, the "Elephant Man" who claimed that his deformity was caused by his mother being frightened by an elephant during pregnancy. It isn't totally clear what Merrick actually suffered from, but the theory is that he was the victim of two rare conditions, neurofibromatosis and Proteus syndrome.

To observe Mary more closely, St. André arranged for her to be taken to London, where other doctors could be privy to the remarkable phenomenon. Alas, no more rabbits were produced. Mary's story began to unravel when a porter was caught trying to sneak a rabbit into her room, explaining that he had been hired by Mary's sister-in-law. That prompted an official investigation, but Mary admitted nothing. Only when she was threatened with surgery to explore her insides did she finally confess that she had inserted the dead rabbits into her birth canal manually and managed to squeeze them out, making it appear as if she were giving birth. Why did she do it? The age-old answer, money. In the eighteenth century it was common for people to pay to see human curiosities, and what could be more curious than a woman who had given birth to rabbits?

Mary spent a few months in jail and then returned to relative obscurity. St. André attempted to vindicate himself by taking out an ad in a newspaper apologizing for his "mistakes" and expressing hope that "people would be able to separate the innocent from those who have been guilty actors of this fraud." But the public was not forgiving. St. André's patients deserted him and he eventually died a poor man.

Cartoonists had a field day with drawings of Mary spewing out rabbits surrounded by caricatures of gullible physicians who had swallowed her story. And that really is what makes this case so fascinating. Although medical education was still rather primitive at the time, there was certainly enough known about anatomy and reproduction to have dismissed Mary's rabbit births as claptrap. But it seems education wasn't then, and isn't now, a vaccination against folly. There are physicians today who advocate against vaccination and many who buy into homeopathy, coffee enemas, antineoplastons, alkaline water, energy healing, and other forms of woo that make no more sense than a woman giving birth to rabbits.

RUDOLF STEINER AND ANTHROPOSOPHY

I had never eaten a biodynamically grown tomato. In fact I had never even heard of such a thing. But the lady in the health food store in Vancouver assured me that this tomato was not only "free of chemicals" but that it had been grown in harmony with the moon and the planets. I long ago learned that it is quite fruitless to get into discussions with the scientifically challenged, so I bit my tongue and, sensing a good story, decided to invest a small fortune in the biodynamic tomato. Storywise, it turned out to be a good investment. Little did I dream that a little research into biodynamic farming would lead me to the strange world of anthroposophy and its mystical founder, Rudolf Steiner!

Steiner was born in 1861 in a part of the Austro-Hungarian Empire that would later become Yugoslavia. By training he was an architect and some of his unusual free-flow structures still stand in Germany and Scandinavia. But it was along quite different lines that Steiner achieved fame, or notoriety, depending on one's views. In 1899 Steiner wholeheartedly embraced a fledgling religion known as theosophy. This had been founded by a Ukrainian "psychic," Helena Petrovna Blavatsky, and was based on a curious blend of astrology, spiritualism, and Eastern mysticism. Blavatsky claimed to be in contact with spirits who would send her written messages that mysteriously floated from the ceiling during séances. On numerous occasions she was caught with conjuring equipment and her spirit manifestations exposed.

Steiner eventually split with Blavatsky, only to found his own peculiar religion, which he called anthroposophy. The name derives from the Greek words for "man" and "wisdom," but the general tenets of anthroposophy are virtually impossible to describe. Steiner certainly claimed to be clairvoyant and his writings speak of the importance of "bringing oneself into harmony with the divine creative force" and "accessing a cosmic reservoir in the astral plane where every thought or action that has ever occurred is recorded."

The esoteric Steiner emphasized the oneness of the spiritual and material worlds and the importance of balancing cosmic forces. Such lingo is not unusual for metaphysical movements, but anthroposophy goes beyond mystical exhortations. Steiner had some, let us say, unusual ideas about how the spiritual and the material world are linked. The soil, he maintained, was a living organism that possessed a life force that had to be constantly replenished. This was not possible with agricultural chemicals, which were in fact destroying the soil. Special composts and sprays were needed to nourish the soil and make it more "spiritually balanced."

So far this just sounds like an odd rationale for the application of fertilizer. But actually, Steiner's methods do not make use of any scientific principles. "Biodynamic farming," a term he coined, requires the use of certain "biodynamic solutions." One of these solutions is prepared by packing cow manure into a cow's horn and burying it for several months. The horn is then dug up and the powder inside mixed with twelve gallons of rainwater in a ritualistic fashion. First the solution has to be stirred for twenty seconds in one direction, then for twenty seconds in the other to "bring the cosmic life forces of the earth and the universe into harmony." This mystical mix is said to "rejuvenate" four acres of soil. Chemically speaking, by this time the manure would be so diluted that it couldn't possibly have any effect. But then again, anthroposophists insist that fertility of the soil has nothing to do with such mundane things as chemicals.

Once the soil has been prepared, crops can be planted. Abiding by the position of the moon and the planets, of course. Steiner believed that the gravitational pull of the moon was critical in the nurturing of plants, and new crops were to be planted two days before full moon. Of course the phases of the moon are not linked to its gravitational effect on the Earth. Planetary positions also supposedly determine when different parts of a plant grow. Lettuce and spinach were to be tended on "leaf" days and potatoes on "root" days. Steiner also believed that the flapping of birds' wings and their chirping influenced plant growth. The right frequencies spurred the plant's development, he maintained.

Biodynamic farming is still practiced in several countries, sometimes even using a concert of birdsongs. Claims are becoming even more elaborate. In fact, a couple of years ago the Bio Dynamic Farming and Gardening Association of New Zealand offered to come to the aid of the government. Possums, which had been introduced into New Zealand from Australia with the aim of establishing a fur industry, were instead destroying forests by eating everything in sight. Neither trapping nor poisons were able to make a dent in their population. But the biodynamic farmers had a solution!

Possum testicles were to be burned and mixed with sand to make "possum pepper." Then a highly diluted extract of this concoction was to be spread around the possums' habitat. The possums would flee in terror, it was said, from a mere whiff of the emasculating mixture. The desperate New Zealand Institute of Forestry actually carried out a double blind, placebo-controlled study of this bizarre technique on both wild and penned possums. Biodynamic materials had absolutely no effect. The possums did not even play possum.

Anthroposophy also has a unique approach to medicine. The body has three poles, identified as cool, warm, and balancing. Illness comes from a disharmony of these poles and can be restored using a variety of animal, mineral, and plant substances. The time of day and planet constellations have to be considered for the preparation of these remedies and they must never be prepared between noon and 3:00 P.M. because this is the "least alive" time of day. Vaccinations are frowned upon, while color therapy as well as a mistletoe preparation invented by Steiner are promoted. Sauerkraut is regarded as a special dietary food required for the health of the digestive tract.

To a scientist, this of course is silly stuff. Even Steiner admitted this. "I know perfectly well that all of this may seem utterly mad," he once said. "I only ask that you remember how many things have seemed utterly mad which have nonetheless been introduced a few years later." An interesting notion, but quite misleading. The fact is that most ideas that seem utterly mad, are utterly mad. But at least in one instance, Rudolf Steiner's views transcend the metaphysical drivel. He states in one of his books that: "If the blond and blue eyed people died out, the human race will become increasingly dense if men do not arrive at a form of intelligence that is independent of blondness." Scary stuff.

Anyway, I now know what biodynamic farming is all about. And what did my tomato taste like? Basically like any other tomato. Maybe the birds didn't sing enough to it. Or maybe the planets were not properly aligned when it was planted. The saleslady agreed that the tomato was not as good as usual. Perhaps the grower had not done everything required. Perhaps he had neglected to use "earth acupuncture." Don't even ask.

PERKINS TRACTORS

"Where's the evidence?" That's all Dr. John Haygarth wanted to know. The English physician, who had retired from practice in 1798, had not lost his interest in medicine and was troubled by the Perkinism that was sweeping England. Haygarth was skeptical about the plethora of medical problems supposedly solved by the simple act of drawing the tips of two metallic rods along the sufferer's skin. Of course, these were no ordinary metallic rods. They were Dr. Elisha Perkins's metallic Tractors, made of a special alloy, rounded at one end, pointed at the other, and designed to rid the body of "the noxious electrical fluid that lay at the root of suffering." Dr. Haygarth had a different prescription for the noxious electrical fluid. He recommended a healthy dose of debunking.

Back in the eighteenth century medical education was a haphazard business, with the trade usually learned through apprenticeship. Young Elisha had learned doctoring from his father, Dr. Joseph Perkins, and had succeeded in setting up an extensive practice in Plainfield, Connecticut. It was here that he made his amazing discovery. That is, if it really was his.

According to Elisha's account, one day while performing surgery, he noted the contraction of a muscle on contact with the point of one of his metallic instruments. This was a curiously similar observation to one that Luigi Galvani had made in Italy a couple of years earlier with his famous frog legs. Galvani had shown that the detached legs quivered as if alive when probed with a pair of dissimilar metallic rods. He misinterpreted the phenomenon, ascribing it to the release of "animal electricity." His countryman Alessandro Volta correctly interpreted the experiment, explaining that two dissimilar metals were capable of generating a current when connected through an appropriate medium, in this case the fluid in the frog's leg. Building on Galvani's observation, Volta introduced what came to be known as the voltaic pile. Constructed of alternating discs of zinc and copper, separated by pieces of cardboard soaked in brine, the pile was the first practical method of generating electricity. Volta had made the world's first battery and, in honor of Galvani, coined the term "galvanism" for the electrochemical process that made it possible.

Whether Perkins was aware of Galvani's experiment, or made a similar observation independently, isn't clear. But he certainly did use the term "galvanism" in the commercial promotion of his Tractors. So how did a pair of metallic rods become miraculous curing agents?

After noting the peculiar muscle contractions, Perkins began to wonder about the role of the metal instruments in the effect. Was the critical feature the material of which they were made? Or was it possible that any similarly shaped object would do the same thing? As it turned out, that was not the case. Unable to reproduce the contractions with instruments made of wood, Perkins concluded that metals somehow had the ability to affect tissues. His curiosity aroused, he began to investigate the effects of instruments made of various metals and found that sometimes just resting these on the skin before making an incision eased a patient's pain. When he noted that he could also relieve pain by separating a patient's gum from their tooth with a metal scalpel before an extraction, he was sold on the potential of metallic therapy.

Before long Perkins was experimenting with single rods, as well as with combinations of rods made of various metals. In some mysterious fashion he came to conclude that a pair of rods — one made of copper, zinc, and a little gold and the other made of iron, silver, and platinum — were just what the doctor ordered. When it came to relieving pain, sliding the rods over the affected area exceeded his most ardent expectations. The time had come to let the public, aching for pain relief, in on this therapeutic bonanza. Of course such spectacular relief from pain would not come cheaply. But Perkins maintained that even at their high cost, the Tractors were well worth the investment. And judging by the gushing testimonials, loads of patients agreed!

Perkins was undoubtedly convinced that he had made a wondrous discovery and began to promote the Tractors vigorously to hospitals and to the public. Even George Washington bought a pair. But the

Connecticut Medical Society was not amused, expelling Perkins for "delusive quackery." Maybe the English, Perkins thought, would see things differently! So he sent his son Benjamin to England to present the claims of the healing powers of the Tractors. The Brits were electrified. The Tractors, they said, even worked on animals. So strong was the belief in tractorization that it led to the founding of the Perkinean Institute in Soho with an endowment greater than that of any hospital in London at the time.

Like their American counterparts, British doctors were strongly against the Tractors, claiming that the "cures" were brought about through imagination. Where is the evidence for any other type of activity? asked Dr. John Haygarth. There wasn't any. And Haygarth proved it. He was able to obtain equally wonderful effects with wooden tractors painted to look exactly like the genuine ones. As a follow-up, Dr. Richard Smith of Bristol showed that two common iron nails coated with sealing wax worked just as well, especially if the doctor also held a stopwatch in his hand to further promote the scientific nature of the enterprise. Benjamin Perkins saw the writing on the wall and returned to America while the going was still good with some $50,000 profit.

Perkins's wonderful Tractors have been relegated to history's medical junk heap. At least in their original form. But are not today's therapeutic touch, alkaline water, detoxifying footbaths, and healing bracelets just their reincarnation?

IRIDOLOGY AND CRANIOSACRAL THERAPY

Let me tell you a story. Sometime in the 1830s, Ignaz von Peczely, a young Hungarian schoolboy, was attacked by an owl. A fight ensued and Ignaz managed not only to fend off the attacker but to break its leg. Probably feeling somewhat remorseful, he took the bird home, hoping to restore it to health. Looking into the eye of his captive, the boy noted a dark stripe in the iris, the colored part of the eye that surrounds the pupil. This, the ingenious student surmised, must be a consequence of the broken leg. As the leg healed, Ignaz claimed the black stripe eventually became a tiny black spot surrounded by white lines. The iris, he suggested, was therefore a window into the health of the body. This curious tale, believe it or not, gave birth to the bizarre practice of iris diagnosis, or iridology.

Peczely went on to become a physician and apparently spent a lot of time staring into the eyes of his patients. He produced numerous charts linking the properties of the iris to various organs and claimed that patterns in the iris mirrored "imbalances" in the body. This puffery was popularized in the U.S. by Bernard Jensen, a chiropractor who produced books and charts of all kinds aimed at diagnosing medical conditions by examining the iris. What he didn't produce, though, was any evidence that this silliness had any merit. Neither has anyone else come up with any supporting data. But this certainly is not for lack of trying. In a classic case, Jensen and two other iridologists were given photographs of 143 people, 48 of whom had been diagnosed with kidney ailments. Could they identify the kidney patients? The experiment was chosen because iridologists routinely claim to be able to do this. All three failed miserably.

Numerous other such studies have had the same outcome. At the University of California at Berkeley, iridologists were shown color slides of the irises of thirty patients who had some sort of orthopedic trauma to the leg or arm, along with pictures of the eyes of thirty matched subjects with no history of such trauma. Although many iridologists were asked to take part in the study, only three volunteered. Surprise, surprise! By not taking part, they avoided the embarrassment that the three volunteers experienced. They were totally unable to match the irises to the broken bones. So, just how many failed experiments do we need before we can pile iridology on the medical junk heap?

I can't say that iridologists are mired in the past though. Some now use a computerized instrument (Bexel IRINA) that scans a patient's iris to diagnose ailments. What sort of ailments? According to some

Korean "studies," anything from digestive problems to cancer. This is not benign nonsense. A clean bill of health from an iridologist may prevent diagnosis of an existing ailment, and some sort of false diagnosis can scare people unnecessarily. One iridologist, for example, diagnosed a patient with "toxic environment in the digestive tract." The recommendation was "herbal cleansing of the kidney and liver," whatever that may mean. But what can you expect from a practice based on a young boy looking into the eyes of an injured owl? Iridology is for the birds! So, I think, is craniosacral therapy.

This rather unusual regimen can be traced back to Dr. William Sutherland, an American osteopath who practiced in the first half of the last century. Osteopaths believe that physical manipulation of the skeleton can alleviate many health problems. But Dr. Sutherland added a further twist. He contended that manipulating the bones of the skull was the key to curing illness. Why? Because such manipulation would affect the functioning of the cerebrospinal fluid, the fluid that surrounds the brain and the spinal cord. Sutherland noted that this fluid pulsed rhythmically and somehow concluded that changes in the natural rhythm caused disease. These irregular pulsations could be corrected, he maintained, by gently manipulating the bones of the skull in order to alleviate restrictions on the flow of the cerebrospinal fluid.

Sutherland was promptly labeled a heretic and a quack by other physicians but received strong support from many patients who claimed that a variety of health problems resolved with craniosacral therapy. And what does modern medicine say about this? Pretty well that it's all bunk. The bones of the skull are not amenable to manipulation as Sutherland and his later followers claim. They actually fuse during infancy. While it is true that the cerebrospinal fluid does pulse, this is actually related to blood flow, not to any mysterious force. Indeed the whole idea of a craniosacral rhythm cannot be scientifically supported.

When different practitioners are asked to measure this supposed rhythm by placing their fingers on a patient's head, they come up with vastly different craniosacral rates. This isn't surprising, since they are trying to measure something that doesn't exist. Then in response to their measurements, they apply specific manipulations to the skull and claim to be able to help chronic back pain, autism, asthma, learning difficulties, fibromyalgia, and a host of other conditions. Practitioners also claim that their skull manipulations are preventative and can bolster resistance to disease. They report that patients who have regular craniosacral adjustments feel more energetic and happier.

The greatest proponent of this therapy was Floridian osteopath Dr. John Upledger, who passed away in 2012. When his skull manipulation didn't work, he had other approaches. For example, an overanxious patient, diagnosed by Upledger as suffering from excess energy, would have the excess drained out by connecting the patient's toe to a drainpipe with a copper wire. In one case, a lady was tethered with a thirty-foot wire so she could still whirl around the house as her energy was drained away. What can I say? She should have her head examined.

ALKALINE NONSENSE

It is not often that I'm left speechless. But sometimes you run into a situation where words just fail you. Absurd, ridiculous, ludicrous, preposterous, comical, and farcical come to mind, but they still don't quite seem to capture the extent of the mind-numbing nonsense. And what nonsense is that? Ionized Alkaline Water! People, seduced by the outlandish promotional drivel, are spending thousands of dollars for a device that produces this liquid malarkey.

Some promoters just blather mindlessly about increasing energy, reducing weight, reversing aging, boosting immunity, controlling blood pressure, cleansing the colon, or eliminating body odor. More disturbing are the ones who speak of preventing cancer and increasing life expectancy. And how is alkalized water supposed to accomplish these miracles?

Well, you see, "all electrons in water either spin to the left or the right and high speed of the left spin of electrons is considered to substantiate that the water is vital and alive. Only ionized water has this quality," claims one marketer. Uh huh. There's more. "Ionized water oxygenates the body via an increase in the oxygen-hydrogen angle. All other water is void of this benefit." Yeah, sure. "Ionized water has positive polarity. Almost all other waters are negative in their polarity. Only positive polarity can efficiently flush out toxins and poisons in the body at the cellular level." There's still more. The amazing water ionizer produces "smaller water molecule clusters which enables every nook and cranny of your body to be super-hydrated." Makes your head swim.

All this rubbish does have an effect. It makes anyone with a chemistry background want to tear their hair out. Of course, the promoters of ionized alkalized water have an answer to that too. They claim the water has a calming effect and can even grow hair. Not only is there not an iota of scientific evidence for any of the claims, the notion of ionized alkaline water having any therapeutic effect is beyond absurd. In fact, the term "ionized alkaline water" is scientifically meaningless.

What then does an ionizer actually do? The same thing that high school students do in chemistry labs when they stick a couple of electrodes in water and pass a current between them in a classic electrolysis experiment. Some of the water molecules break down, forming hydrogen gas at the negative electrode and oxygen at the positive electrode. Electrolysis, however, cannot be carried out with pure water since water cannot conduct an electric current. For electrolysis to proceed, some sort of charged species must be dissolved in the water. Atoms, or groups of atoms that bear a charge, are called ions. Tap water contains a variety of dissolved ions such as calcium, magnesium, sodium, bicarbonate, or chloride that make it amenable to electrolysis.

As water molecules break down at the negative electrode to release hydrogen gas, they leave behind negative hydroxide ions. This is what makes a solution alkaline. Basically what this means is that as electrolysis proceeds, a dilute solution of sodium hydroxide (negative ions are always paired with positive ones) is produced around the negative electrode and can be drawn off as "alkaline" or "ionized" water. But you don't need an exorbitantly expensive device to produce a dilute sodium hydroxide solution. A couple of pellets of drain cleaner in a liter of water will do the job. So will a spoonful of baking soda. Of course these solutions will not produce any medical miracles. But neither will the posh alkaline water.

What this expensive water does produce is a bevy of daft claims. Here is the most popular one: "It is well known in the medical community that an overly acidic body is the root of many common diseases, such as obesity, osteoporosis, diabetes, high blood pressure and more." Poppycock! There is no such thing as an acidic body. That, though, doesn't stop the hucksters from treating it. How? By neutralizing the acidity with their alkaline water. "The alkaline water will restore your body to a healthy alkaline state," they say. "It counteracts the acidic food you eat and the effects of the harsh elements in your environment in order to bring about the natural balance your body needs. Change your water and change your life." The only thing you'll change is your bank balance.

Now, even if there were such a thing as an acidic body, and even if this signaled illness, it could not be countered by drinking alkaline water. To "alkalize the body," one would have to alkalize the blood. But our body maintains the pH of the blood between 7 and 7.4, which is already alkaline. If you were to alkalize it further, you would not have to worry about illness because you would be dead. Don't worry, though, about alkaline water killing you. Our stomach is strongly acidic and any base that enters is immediately neutralized. The still-acidic contents of the stomach then pass into the intestine where they are neutralized by alkaline secretions from the pancreas. So all of the water we drink ends up being alkaline anyway!

Another seductive claim is that alkaline ionized water is an antioxidant and neutralizes free radicals. This is often demonstrated by immersing an oxidation-reduction potential (ORP) probe into the water and pointing out that the needle moves into the negative millivolt region, while ordinary water shows a positive reading. An ORP probe is useful for determining water quality in a swimming pool, but is meaningless for drinking water. The slightest amount of dissolved hydrogen, as you have in alkalized water, will result in a negative reading. This has absolutely no relevance to any effect on the body. Oil may not mix with water, but it seems snake oil surely does.

SPOON-BENDING FIASCO

Everyone has skeletons in their closet. There's at least one in mine. A couple of years ago while on a cruise I pinched a spoon from the dining room. It wasn't because of any lack of spoons at home, it was because no matter how hard I tried I could not bend this one. I tried with two hands, I tried by pushing against the table, I even tried placing the handle under my heel and tugging on the head. No give at all. I had to have that spoon!

I've been practicing magic as a hobby ever since I was a teenager. It has turned out to be a perfect fit with my career because of the numerous scientific principles involved in creating the illusion of contravening the laws of nature. And that is what magic is all about. Seeing someone levitate, or vanish inside a cabinet, or appear out of thin air requires an apparent suspension of the laws of nature. The key word of course is "apparent," because all such effects are accomplished by clever scientific means. A magician, however, attempts to ensure that the audience will not discover those means. Science can also appear magical, but in this case, we relish in scuttling the magic with down-to-earth explanations. Just think about it. Isn't an airplane with hundreds of people aboard flying through the air magical? How about taking pictures with your smartphone and sending them around the world in seconds? Or a seed growing into a plant or a new life being created from the meeting of cells? But magic is converted into science with an appropriate explanation.

I have found performing magic to be an excellent springboard for a discussion of scientific methodology and for fostering the critical thinking needed to prevent being swept away by the tsunami of pseudoscience generated by a rapidly multiplying bevy of charlatans. When you can demonstrate how "psychic surgery," a procedure by which diseased tissues are apparently removed without an incision, can actually be accomplished by sleight of hand, you have given believers something to think about. Similarly, a demonstration of "mental" effects with a clear declaration that these are done by clever chicanery can help convince at least some that trickery may be involved when psychics perform feats that seem scientifically inexplicable.

One such feat is "psychokinesis," or the ability to move objects using only the power of the mind. Psychokinetic effects were first popularized in the middle of the nineteenth century when Angelique Cottin in France claimed that electric emanations from her body allowed her to move objects without touching them. She convinced many observers of her power, but critics offered quite simple explanations about how such effects could be performed by natural means. Since that time numerous psychics have claimed psychokinetic powers, with Uri Geller being perhaps the most famous. In the 1970s, he beguiled audiences and even some scientists with his apparent ability to bend metal with the power of his mind. He gets credit for introducing the phenomenon of mental spoon bending, an effect upon which he built quite a spectacular career.

Magicians were also astounded. Not by the effect, which can be accomplished by a number of established methods, but by how the public was so ready to swallow a paranormal explanation. Conjurers were quick to reproduce the spoon-bending trick, pointing out that the only requirement was a modicum of sleight of hand. This brings us back to my pilfered spoon.

When I do the spoon-bending trick, I first hand out the spoon to the audience with a challenge to bend it. Once it is established that it can withstand all efforts, I proceed to bend it "with the power of my mind." But in rare cases, some strong men have managed to bend the spoon and destroy my performance, so I'm always on the lookout for super-strong spoons. I can tell you that Crystal Cruises have such. They absolutely cannot be bent, except in the hands of a magician who is equipped with a "special something."

But why am I talking about tormenting cutlery? Because back in 2021, thanks to colleague Tim Caulfield, a health law professor at the University of Alberta, I learned that Integrative Pediatric Medicine Rounds at his university were set to feature a talk on "Spoon Bending and the Power of the Mind." The seminar would be given by an "energy healer" who was described as being "a Reiki Master teacher, a certified Trilotherapy practitioner, a Yuen Method practitioner and a teacher of popular Spoon Bending and Tantric Sex workshops." So this was not to be a workshop on critical thinking, which could have been appropriate. The prospective speaker actually claimed that 75 percent of attendees would be able to bend spoons with their mental energy!

The scientific community reacted with vigor to this assault on reason, and the resulting extensive media coverage caused the seminar to be canceled, with some weasel explanations being provided about the workshop "being withdrawn by the presenters."

The presenter was to be Anastasia Kutt, who was not some wacky outsider but rather was listed in the university's directory as "a research assistant in the 'Complementary and Alternative Research and Education (CARE) Program'" who "is also involved in research activities and organizing events." What sort of events? Given her interest in topics such as tantric sex and spoon bending, one wonders.

Criticism of this spoon-bending fiasco should not be construed as an attempt by the mainstream scientific community to curb free speech or to police academic research. Rather it is an appeal for reason and for vigilance against quackery sneaking into integrative medicine programs, which are becoming increasingly popular.

I don't know how Ms. Kutt bends spoons, but I was willing to fly to Edmonton at my expense to find out. I was even ready to eat a University of Alberta Integrative Health Program hat were she able to bend my Crystal Cruises spoon. I think my dining habits would not have been altered. In any case, I was happy to learn that she is no longer associated with the university.

QUACK PRODUCTS

I have to tip my hat to the algorithms that Google and Amazon use to detect people's interests and then target them with specific ads. Since I often carry out searches that use terms like "alternative medicine," "naturopathy," and "water treatment," I am inundated with ads for products that seem to fit these categories. I thought it would be fun to make a collection of some of the more interesting products that were pitched to me over one week.

There were magnetic toe rings for weight loss, healing crystals from India, acupuncture point–stimulating Chinese balls, and detoxifying Tibetan singing bowls. But the one that really caught my attention was the Belly Button Tool, designed to "push the greatest button you will ever push." What is it? A wooden rod with a silicone tip that fits into the belly button (not sure what to do if you have an outie). How do you use it? "Rapid, gentle, in and out movements, going clockwise for 5 minutes when you wake up and 5 minutes before you go to sleep."

Why? Because this manipulation will lead to "weight loss, increased energy, relief from stress and anxiety, relief from pain, an increase in flexibility by 25% and a raised immune system." A "loosened gut" to boot. How does tormenting your belly button in this fashion accomplish these wonders? According to the "Doctor of Detox" who markets the gizmo, it stimulates the vagus nerve, which is desirable since "vagus nerve dysfunction is responsible for numerous ailments." He should know because he is a "medical intuitive," a Doctor of Natural Medicine, and a Doctor of Humanitarian Services. Not sure why a finger would not work just as well.

This outstanding humanitarian is also concerned about the water we drink. We learn that "as water travels the earth it becomes symmetrically structured in a sturdy and elaborate geometric shape, reducing its surface tension, neutralizing toxins and increasing its hydration power." Unfortunately, water treatment "wipes the memory of water clean." Luckily for us, this icon of science knows how to remedy the situation. He sells devices that can restructure the water. "Structured Water creates a greater flow of energy as it connects with life and when taken into our body, it greatly enhances our body's ability to rejuvenate and function in a more optimal manner." A portable gold-plated Structured Water device will set you back $1,495, but if you can forgo the gold, you can avail yourself of this miracle for $349.

Another ad prompted me to purchase Dr. Hidemitsu Hayashi's Original Hydrogen Rich Water Stick. This is to be immersed in water to hydrogenate it and "increase health and vitality." There is no mention of how this is accomplished. The stick turned out to contain bits of magnesium that can indeed react with water to produce hydrogen gas, although I did not note any bubbles when I followed the instructions. I should point out that hydrogen gas is virtually insoluble in water and any that does dissolve would quickly outgas. Nevertheless, I did as I was told and consumed the "hydrogenated water" for a week. Not even a burp.

Next, I ordered the Kikar Portable Electric Hydrogen Water Ionizer. This promised not only to add hydrogen to water, but to "convert it to the hydrogen anion which is a powerful antioxidant because it scavenges reactive oxygen species." The device turned out to be a bottle with electrodes at the bottom that when connected to electricity carried out electrolysis. This time bubbles of hydrogen did indeed form, but the notion of converting hydrogen to anions through electrolysis is total nonsense. The literature with the product also claims that "hydrogen has been found to have anti-tumour effects," hoping to snare some desperate people.

By now I was getting kind of stressed, so I turned to another invention that promised relief, the Quartz Crystal Elixir Water Bottle. Drinking from it would "boost my spirits, gently neutralize negative vibrations, bring emotional calmness and relieve stress and anxiety." Surely this would work, since it was backed by that shining star of science, Gwyneth Paltrow. The bottle contained an attractive large crystal, the properties of which were said to be transferred to the water "through molecular vibrations, charging it with energy, soothing our mind and emotions." The only emotion it generated in me was irritation.

The same can be said for PolarAid, a "revolutionary, easy to use, portable, handheld body tool that supports every aspect of health by harnessing the vital frequencies that naturally surround us at all times." True, it is easy to use. It is nothing but a plastic disk that looks like a coaster. It can even be used as such. We are told that "a glass of water becomes energized when placed on it, and inserting it under a potted plant makes it thrive." For health, PolarAid can be used to "unblock blocked energy channels by holding it over chakra points." Once the channels are unblocked, health is ensured by sitting on PolarAid for thirty minutes a day. All this gives me indigestion. I think I may need a belly button massage.

BELIEVING THE UNBELIEVABLE

"You're just a skeptic" was the accusation when I ventured the opinion that "no-touch knockout" had nothing to do with any mythical qi, but had plenty to do with the power of suggestion. I had just given a talk on the power of the mind, mostly focusing on placebo and nocebo effects, when I was asked by a member of the audience why

I had not mentioned the amazing ability of qigong masters to focus their qi and knock down a person without touching them. It was when I questioned the existence of such a form of energy that I was called a skeptic in somewhat of a deprecating tone. Indeed, I fully admit to questioning the validity of claims purporting to be factual, in other words, to being a skeptic. A skeptic is not a person who refuses to believe, just one who says show me the evidence before I hop on the bandwagon.

I am certainly not hopping on the "no-touch" bandwagon, but I have seen it roll by. I have watched videos in which martial arts experts knock down people without touching them, seemingly with some invisible force field emanating from their hands. There has never been any scientific evidence of such a field, but that doesn't mean it is all a scam. People can really be knocked down, the criterion being that they believe qi exists, and that it can be discharged by someone who has mastered the art. What we are looking at here is the power of suggestion, which is amazing in its own right. If you have ever witnessed a hypnosis stage show, you may have seen a subject drop an object they were given to hold upon being told that it is becoming burning hot. Of course, there is no change in temperature, but the subject is totally convinced that there is. Similarly, if you believe that someone has the power to knock you down without touching you, possibly because of having seen this happen to others, you can indeed have a similar experience.

I'll admit that the videos can be very persuasive, but the dictum that "seeing is believing" is not to be believed without further exploration. In this case, that has been done, in quite an amusing fashion. One of these "qi masters" challenged a karate expert to a match, claiming that he would triumph without touching his opponent. It took only a few seconds before he ended up in a bloodied heap on the floor. An even more comical situation was captured in a segment of National Geographic Channel's *Is It Real?* program when a "qi master" trained by karate expert George Dillman, the most famous of the "no-touch" proponents, tried in vain to knock down skeptical investigator Luigi Garlaschelli. Dillman had an explanation for the failure: "If the guy had his tongue in the wrong position in the mouth, that can nullify qi power. You can also nullify it by raising one toe and pushing the other one down." This was said in all seriousness. Quite humorous.

There are many other claims of supposed paranormal powers that are not quite so humorous. I have a long-standing interest in such phenomena, sparked back in 1975 by a demonstration of "psychic surgery." I was already intrigued by magic at the time and was excited to see an ad for a course offered by Henry Gordon, a well-known Montreal magician. The ten sessions, given in his magic shop in the evening, introduced us to the various aspects of the art, from close-up to stage illusions. Henry, like most magicians, was a devotee of critical thinking and was very annoyed by people who used magic tricks to pretend they had real powers.

It was in this context that we learned about "psychic surgeons" in the Philippines and South America who seemingly were able to remove tumors from patients' bodies without making an incision. Henry went on to demonstrate the stunt in front of us. I was in the first row and was thoroughly captivated. What looked like a piece of bloody tissue was removed from the belly of a student who had volunteered to be the guinea pig for the demonstration. I couldn't figure out how this apparent miracle had been accomplished, but since this was a course in magic, we would find out! And we did. Once we learned about the "special something" that was needed, we too could perform the miracle. Was that ever an eye-opener! A phenomenon that seemed to defy the laws of nature was no more than a magic trick. Albeit a very convincing one.

Ever since then, I have been an avowed skeptic and have tried, when possible, to deliver a knockout to pseudoscientific claims such as the no-touch knockout.

ENCOUNTERING A "HEALER"

The mountain in the Sinai desert beckoned mysteriously that early morning. Urged on by an inner sensation, he set out alone towards the top. And when he reached it, something wondrous happened. Suddenly he was enveloped in a brilliant light that left him a changed man, a man now gifted with a power to help humanity. Moses over 5,000 years ago? No. Ze'ev Kolman, an Israeli army reservist, in 1974.

According to Kolman, as he gazed out from the summit he noticed an elliptical, bean-shaped object quickly pass above him. The next thing he knew, he was in the center of some sort of cloud, surrounded by eleven hairless human-like creatures, speaking in a language he could not understand. Then they all vanished, leaving Kolman lying on the mountain top, confused.

The next surprise came when he returned to the army camp and brushed by a fellow soldier who collapsed as if struck by lightning! But when he came to, he felt "fantastic." Other soldiers wanted to experience this amazing effect, and Kolman complied. A wondrous energy seemed to emanate from his hands, which not only made the soldiers feel elated, but made lesions on the knees of one disappear, and cured another of chronic headaches. Kolman realized that somehow, by some higher power, he had been given the gift of healing. It would now be his duty to use it to help mankind. Parapsychologist Hans Holzer in his book about Kolman, *The Secret of Healing*, opines that Kolman's encounter was probably with extraterrestrials, who were carrying out an experiment to see how changing a person like Ze'ev through their advanced technology would play out in the field of human beings.

It was not only extraterrestrials who may have visited Mr. Kolman; he's had encounters with dead people as well. He tells of a meeting with a holy man named Baba Salim two years after his death. Baba appeared one night in front of Kolman and placed an empty bottle into the folds of his blanket, saying "take the bottle, fill it with water, and give it to your sick patient — he will recover." Kolman had other visions too. He began to notice auras around people, and from their shape and color, he was able to determine the person's health status. He then found that he could manipulate these auras with the "bioenergy" projecting from his hands and improve their health.

On occasion, the auras even allowed Kolman to look into a patient's past, once determining that a man's neck pain came from a previous incarnation in which he had been hanged from a tree. In another instance Kolman was able to expel the spirit of a dead woman who had entered a bride's body when she had relations with her husband for the first time. His main talent, though, seems to be in healing people. With bioenergy flowing from his body, at least according to testimonials, he has improved the lot of celebrities such as Carly Simon and Robert Wagner, various politicians, and numerous others. There are accounts of reducing back pain, treating eye, ear, and breathing problems, and even curing cancer. Kolman does not claim any sort of medical expertise and advises people to see physicians, although he does have some interesting opinions. He believes that 80 percent of all cancers stem from emotional causes and that the success of bioenergetic healing is higher if it is attempted before chemotherapy.

Surprisingly, in spite of Mr. Kolman's apparent exhibition of astounding powers over some thirty years, I had not heard of him until 2006 when a friend called to inform me that he was coming to Montreal for healing sessions, to be preceded by a public lecture. This prompted me to look at Kolman's website and to discuss the information he offered there on my radio show. I expressed some skeptical comments, particularly about the spring water he was selling, "energized by Ze'ev and powerful in healing all kinds of ailments." All one had to do was rub three drops into the forearm three times a day. A bargain at $35.

My comments seemed unfair to the gentleman who had organized the visit and prompted an invitation to meet Kolman so that I could witness his abilities myself. Expecting the dashing, debonair man of the publicity pictures, I must admit I didn't recognize the paunchy, jovial fellow who walked into the room. We had an interesting conversation about his abilities, and he offered me a demonstration if I would return the next day. He would waive the usual $250 fee he charged for a session. As you can imagine, I didn't hesitate to accept.

Kolman was pleasant enough and told me that he didn't really understand why he was able to do the things he did, but he knew that his patients told him that they could feel the "bioenergy" emanating from his hands, and that he had all sorts of testimonials indicating that they had been helped. In some cases, he could even work his magic over distance. All he needed was a picture of the patient and he could then dispense his energy appropriately. I asked if he could demonstrate this remarkable ability. "Sure," was the reply, whereupon I handed over a picture of a colleague who Kolman diagnosed as having some sort of problem. Given the situation, not a surprising diagnosis. Precisely at 11 P.M. he would transmit his energy and attempt to heal him. I should alert my colleague to this, so that he could document what happened. Alas, nothing did.

In any case, Mr. Kolman told me that he had numerous testimonials from people who had been helped with diseases ranging from skin and eye problems to heart disease and cancer. All of these appeared to respond to Mr. Kolman's hand waving and energy emissions. I too could experience the force, he told me, if I wished. Indeed, I wished.

When I returned the next day, as I had been asked to do, Kolman examined my aura and offered to rebalance it, which I gladly accepted. After removing my shirt, I lay down on a cot as the healer hovered his hands over my body. Suddenly I felt the energy! No doubt about it! Something was going on. It was as if a wind of tiny dust particles, or perhaps little needles, was sweeping over my body. Exactly as had been described by a number of people I had talked to who had attended Kolman's public performance, during which he had walked among the audience, hand extended, letting the energy flow. Indeed, this had raised my curiosity, because although imagination can do some amazing things, there were too many people describing a similar effect. When my assistant, who Kolman had graciously offered to treat the day before for free, also described the sensation of a strange wind over her body, I knew the effect was real. She was scientifically trained and not prone to letting her imagination run wild. And now I felt Kolman's power too! But I wasn't all that surprised. That's because the night before I had sought help from a higher authority.

I had called James Randi, skeptic extraordinaire, who ran the James Randi Educational Foundation in Florida. "The Amazing Randi," trained as a stage magician, was an expert at detecting fraud, and had a standing offer of one million dollars to anyone who could demonstrate some paranormal ability under properly controlled conditions set up by experts. I had hardly begun my story when the Amazing One said "Aha!" He proceeded to tell me about various devices that could be attached to the body to generate an electric field. They produced a very high voltage but very low current. The effect is a field around the body that can even be strong enough to light a fluorescent tube in the hand. Indeed, I instantly recalled a youthful adventure at an amusement park in Montreal, when I gawked in awe at a man standing on some sort of box, holding a fluorescent tube, lighting it at will with no wires attached. Randi went on to describe how he had caught a "healer" in Malaysia with a device implanted in his shoe! His antics with bioenergy oozing from his hands, the emotional testimonials, and of course the shoe gimmick were all documented in one of Randi's TV programs.

I was thinking about this as I was lying there on Kolman's cot, experiencing the energy. Actually, this was not the first time I had felt something like this. The sensation was just like the one a person gets when placing a hand in front of an active television screen, or the one experienced during a session with a Van de Graaff generator at a science museum. Was Kolman actually able to generate an electric field? Was he some sort of human version of an electric eel? Or was something else going on? Based upon what I saw and felt, I could not tell. When Mr. Kolman asked if I felt something, I readily admitted that I did, but brought up the topic of high-voltage electrical devices. He said he had never heard of such things, that he was not very sophisticated technically, and that while perhaps effects similar to his could be produced by such means, his healing came from legitimate "bioenergy."

This would be a truly wondrous thing, I said, but being a scientist, I expressed my desire for some solid proof. Could Mr. Kolman come down to McGill University and produce his energy there under controlled conditions? At first he readily agreed, stating that he had been tested at other universities to the satisfaction of the researchers. Since he would be returning to Montreal in August, we could set up a date then, I suggested. Now things began to go askew, with Mr. Kolman stating that this would take time away from his healing, even though I offered to compensate him at his usual rate. Then he said that he would only subject himself to the experiment if I made a large contribution to the organization that had invited him to Montreal. A charitable fellow! Why then not take up Randi's challenge? Produce the bioenergy and pick up a cool million. That would help a whole lot of people!

Mr. Kolman did not return to Montreal and there was no testing, so I still do not know if he is "for real." There is no doubt, though, that many people feel he has helped them. Bioenergy or placebo effect? Who knows? Does it matter?

Finally, my Kolman adventure has had an interesting spin-off. I too can now release energy from my palms, just like bioenergetic healers. But I will freely tell you that the source of my abilities is a device attached to my leg under my pants. The Electric Touch, available from magic dealers, can work miracles. But I plan to use it for entertainment purposes only.

FORCEFUL SOLE SEARCHING

I got my first pair of Florsheim shoes when I was thirteen years old. It seemed an appropriate bar mitzvah present because, after all, I was now becoming a man. I had earned the right to wear an adult shoe.

It squeaked a little as I recall, but it was a very good shoe. I would have never guessed at the time that years later the company would be sued by a consumers' rights group in California, called the Consumer Justice Center, for false advertising and consumer fraud. What's going on here? How can something as simple as a shoe become embroiled in a courtroom drama?

Well, it can if the shoe is a golf shoe and claims to do more than just provide a comfortable barrier between the foot and the ground. Questions naturally arise when there are claims about increased circulation; reduced foot, leg, and back fatigue; pain relief; and improved energy levels. That seems to be quite an accomplishment for a shoe. Ah, but this was no ordinary shoe. Florsheim's MagneForce had magnets built right into it. And therein supposedly lies the magic. But according to the Consumer Justice Center, there is no magic, just some trickery.

Magnets are very popular those days as supposed healing tools. There are magnetic mattresses, pads, bandages, insoles, rings, and bracelets. You can even buy magnetized water. A remarkable website sells magnetic immortality rings that claim to increase life span. The inventor, American Alex Chiu, offers up incomprehensible equations and diagrams to buttress his claims of having solved the problem of aging.

Perhaps I am just not smart enough to understand Chiu's explanations and diagrams because I'm not wearing the immortality rings. You see, they also boost your IQ to 180! I guess Chiu must wear them all the time, because now that he has solved the problem of immortality, he has gone on to other things. He has invented a teleportation machine. He assures us that "he is not one of those stupid morons who doesn't know what he is doing." Why teleportation? Because when we are immortal we will have plenty of leisure time, which we can use to pop up here or there.

Admittedly, magnets can produce fascinating effects. The idea of an invisible force that attracts iron is indeed fascinating. And of course without magnets we would have no electric motors, tape recorders, VCRs, or indeed credit cards to pay for them. But using magnets for healing is another matter altogether. Unfortunately, very scientific-sounding claims about healing abilities can be made and believed by people who do not have a good grasp of magnetism. This, of course, means most people. There is a pattern to these claims.

Usually it all begins with a reference to some form of ancient wisdom. Like how Hippocrates, that most famous of all ancient doctors, used magnets to heal the sick. Or how Cleopatra wore magnetic jewelry to preserve her youth. The fact is that neither the ancient Greeks nor Egyptians ever used magnets in this way. But what if they had? They did many senseless things. Hippocrates, for one, believed that a mixture of horseradish and pigeon droppings could be used to treat baldness. Anyway, after supposedly having established the long and fruitful history of magnetic therapy, the scene often shifts to those flag-bearers of our future, those modern knights, the astronauts. The story is that magnets incorporated into spacesuits resolved many of the astronauts' health problems. And a story it is. No magnets have ever been incorporated into spacesuits for this purpose.

But the real "scientific" selling point revolves around the so-called electromagnetic nature of the human body. There is usually talk of how our nervous system relies on small electric currents and how magnetic resonance imaging (MRI) machines diagnose disease by examining changes in magnetic fields inside the body. Both of these are true. But then from these observations we are asked to conclude that applying small magnets to the body can treat ailments. A scientific and logical non sequitur. First of all, the electricity being talked about really involves the flow of small charged particles called ions. Their motion could in theory be affected by giant magnetic fields but not by the small magnets sold for healing purposes whose strength is in the range of refrigerator magnets. Even magnets used in magnetic resonance imaging, which are orders of magnitude stronger than the healing magnets, do not affect the nervous system and have no effect on blood flow. It isn't surprising that there is no effect on blood flow. While magnet advocates maintain that blood flow is affected because hemoglobin contains iron, the fact is that the iron it contains is not magnetic. And that's lucky, isn't it? We wouldn't want our blood to be ripped out of our body when we're undergoing an MRI scan.

And if magnetic fields can heal, shouldn't we have reports of people being healed, or at least being energized after an MRI scan? The huge magnetic field this instrument generates certainly penetrates the body. Not like the little healing magnets. Those in a shoe produce fields that may penetrate the sock but not much else.

There is another common claim used to buttress magnetic healing. The claim that we suffer from magnetic deficiencies. Physicists say, at least according to the magnet sales people, that the Earth has lost some of its magnetism and since human evolution occurred in higher fields, we are now feeling the ill effects of the reduced magnetism. First of all, physicists do not say this, and even if they did, it would not mean that there is a related health effect. As it is, the Earth's magnetic field varies tremendously. At the poles it is 0.6 gauss, double that at the equator. No one has ever noted any variation in disease patterns based on magnetic field geography. Other related claims are barely worth refuting. Such as the one about the Earth's magnetic core. It goes like this: "The Earth itself is a giant magnet with north and south poles and a liquid core. [True enough.] The hot liquid creates a magnetic field which at the Earth's surface is relatively weak [still true], but serves to keep humans attached to the Earth's surface. Without this magnetic field, we would spin into outer space." This is absurd. As any grade one student knows, it is gravity, not magnetism that keeps our feet firmly planted on the ground.

It would seem then that the arguments used to promote the sales of magnetic healing products are not justifiable scientifically. But that does not rule out their possible effectiveness. Perhaps magnets perform the wonders their advocates claim to have experienced through some completely different mechanism. That's why the only way to study efficacy is through controlled trials. And what do these show? Not much. Although dozens of studies have been carried out, there is only one that is constantly quoted as having shown a positive effect. And that was in a rather rare condition known as post-polio myalgia and has not been reproduced. But trials of magnetic jewelry and even insoles have shown no benefit. Which is exactly why Consumer Justice Center sued Florsheim shoes for making claims that are not scientifically supportable. I wish I could tell you otherwise. I wish that all those people who tell me about their wonderful experiences with magnets were reporting something more than just a placebo effect. Believe me, if any compelling evidence emerges, I'll be happy to relay it. All I can relay now is that the magnetic shoes seem to have disappeared, except on eBay. Available if you are interested in a historic relic of quackery.

EAR CANDLES

"This will really relax you," the charming young "therapist" informed me. Dressed in a crisp white lab coat she seemed really . . . ummm . . . medical. She sounded positively authoritative when she told me that my energy channels would be opened up and that I would be detoxified. Thus enlightened, I was ready for my first ear candling experience.

This little adventure took place at a health fair, basically a trade show for dietary supplements and herbal remedies along with assorted healing devices ranging from crystals and water magnetizers to ear candles. I had read about the latter, but had never seen them in action. Here was my chance! For just $25 I had the opportunity to be candled right then and there. I was going to experience, as the sign above the booth declared, "Thermal Auricular Therapy." And it would be in full view of the dozen or so others who were lining up for their turn to be energized and detoxified. As I waited my turn to lie down on the white-sheeted cot to have a candle inserted in my auricular orifice, I perused the brochure about the treatment I was about to experience. It seems the ancient Greeks were already into candling but it was the Hopi people of Arizona who, as I read, have a long tradition of lying on their side and placing lighted hollow beeswax candles in their ear to improve their health.

And we are not talking about minor health improvements. No sirree. We are talking about promoting lymphatic circulation, improving the immune system, purifying the blood, and opening up chakras. You sure do not want to be running around with closed chakras. Oh yes, ear candling also removes earwax and of course rids the body of those unnamed toxins that cruise through our bodies waiting to wreak havoc.

So, how is this magical process supposed to work? The explanation usually involves reference to the "chimney effect." When a hollow candle is placed in the ear and lit at the end remote from the ear, the hot air rising inside the candle creates a vacuum effect that sucks earwax out of the ear canal. Proof is usually provided to the customer after the procedure by cutting open the remnant of the candle and displaying the brown "earwax" that has been collected. Of course I had a pretty good idea that this was utter nonsense, but still I was eager to have a personal "ears-on" experience. As I lay on my side, the long, hollow candle was inserted in my ear and lit. It took a few minutes for the candle to burn down, a period during which all I felt was a slight warming. Then the therapist took the candle remnant and deftly cut it open, triumphantly displaying the "earwax" inside. There were audible gasps from the amazed onlookers! Did I feel cleansed? she enthusiastically inquired. Well, more like fleeced.

Now it was time for a little demonstration of my own in front of the prospective candlees. I lit my second ear candle (you get two for your $25 investment) without an ear attached. Then I attempted to pick up a tiny piece of tissue paper with the non-burning end. If a vacuum really had been created by the rising hot air inside the hollow candle, this should have been no problem. Of course my attempts were unsuccessful. There is no chimney effect. Just a lot of useless hot air. After the candle had burned down, I duplicated the therapist's technique and cut the remnant open with flair, displaying a load of "earwax" identical to the sample she had supposedly removed from my ear.

Wow, what a miraculous product this was, able to teleport wax from my ear even without any contact! To my astonishment, there were no gasps from the onlookers this time. Perhaps I shouldn't have been surprised by this, given that there were a couple of well-attended booths at the fair manned by healers who were advertising their "remote healing" abilities. All they need to produce a cure is a picture of the patient! By comparison, teleporting earwax seems a simple matter.

I thought that my performance would at least convince some of the victims waiting in line to forgo the candling. Only one actually did! The rest stayed, I suppose seduced by the personal testimonials playing on a television monitor. So, is ear candling just some sort of benign nonsense that provides a placebo effect? No. There are risks.

Hot wax can actually drip into the ear and cause burns and obstructions of the ear canal. A paper published in the respected journal *The Laryngoscope* in 1996 described twenty-one cases of serious injury reported by ear, nose, and throat specialists. In some cases the molten wax actually burned through the eardrum. Such cases, along with Health Canada's own tests showing that ear candles did not live up to the claim of removing earwax, prompted the agency to consider ear candles as a medical device that does not meet safety and efficacy requirements. As a consequence, the sale of ear candles for medical reasons is now illegal both in Canada and the U.S.

That doesn't mean they are unavailable. Hucksters are undeterred by the lack of scientific evidence and have taken to making nebulous claims such as "harmonizing energies." The world's largest producer of Hopi ear candles, Germany's Biosun, does not claim its product removes earwax, it just generates a "massage-like effect" on the eardrum so the user is able to relax, let go, and revitalize like the Hopi. Well, ear candles do not relax the Hopi, who of course have never used such things. Oh, one more thing. At least in two cases, people candled themselves and set their house on fire. One lady died. Quite a price to pay for gullibility. And did I feel relaxed, as advertised, after my candling experience? Nope. Irritated was more like it.

ALPHA SPIN CAN MAKE YOUR HEAD SPIN

I came upon the Alpha Spin device, if you can call it a device, in a roundabout way. I was looking into "Longevity Village" in China, so dubbed because of the unusually long life expectancies of its residents. Proportional to its population, the village has about five times as many centenarians as the rest of China. Located in Bama county, it has become a hot spot for health tourists who come hoping their ailments will respond to the magic of the environment, the same magic that allows the locals to have exceptional longevity.

Some believe the secret is to be found in the air of the area's caverns, naturally enriched with negative oxygen ions. One man claims he beat lung cancer through tai chi exercises in the cave and a diet of boiled pigeons and apples. Others hug giant boulders, convinced that this "geomagnetic therapy" is the key to health. However, most believe that the secret is to be found in the river that winds its way through Bama county. Tourists bathe in the river and consume endless varieties of "longevity water" that are sold everywhere.

It was while searching for research on Bama water that I made the acquaintance of Alpha Spin. Googling "Bama water," the following cropped up: "It is believed that the natural resonance of Alpha Spin is similar to that found in many water springs around the world, including Bama, an internationally recognized longevity village in China. Alpha Spin brings Bama to you by optimizing the natural frequency, stimulating vital life energy, and increasing harmony in body and mind." And down the rabbit hole I went.

I quickly learned that Alpha Spin is a "powerful holistic wellness tool" that "fully optimizes the body's molecular and cellular functions via resonance and the formation of a vortex that results in the expression of a quantum energy field." Furthermore, "the unification of the quantum energy field results in a metatron cube, the result of the interaction of spins allowing the pyramids to communicate by transferring information from quantum energy."

There was more about Alpha Spin's ability to generate hexagonal water clusters, transferring resonance frequencies through water, light, or air, and neutralizing and harmonizing the harmful effects of electromagnetic frequencies. Wow! In all my years of wading through swamps of claptrap I don't think I have come across anything to match the stew of random, garbled, meaningless words cooked up on behalf of Alpha Spin.

Alright, let's put the banal rhetoric aside. What is Alpha Spin supposed to do? We are told that it can energize water and anyone who drinks said water. There is even a way offered to prove this, the "O ring test." The "O" is formed by placing your thumb and forefinger together. Then with your other hand you try to pry the fingers apart and repeat after drinking the "energized" water. The "evidence" is that it is harder to pull the fingers apart the second time because you have been "energized." This is a well-known pseudoscientific ruse based upon expectations. If you believe that it will be harder to pull the fingers apart, then that will indeed often be the case. Of course, when such experiments are done in a blinded fashion, nobody can tell if the water has been "energized."

Alpha Spin also claims to extend the shelf life of fruits and vegetables, improve plant growth, reduce wrinkles, improve circulation, remove energy blockages, treat autism, increase the engine performance of cars, cure cross-eyes, stop jet lag, improve sleep, and, of course, treat cancer. That is because, as everyone knows, "healthy cells have harmonic frequencies that rotate counterclockwise and cancer cells have clockwise spins." And amazingly, Alpha Spin changes the direction of the spin.

As you can imagine, I just had to have this wonder product. It was available on Amazon, so I went for it. Set our office back over $200, but hey, that is a small price for a miracle. I really had no idea what I was getting and waited eagerly for its arrival.

The attractive box said "German Technology" but was mailed from Indonesia. Inside was a little glass plate that looked like a coaster with a hole in the middle. It didn't buzz, didn't spin, had no flashing lights, didn't do anything. It was nothing but a glass disk. For the wondrous effects, you were supposed to hold it over a body part, or a glass of water, or the hose with which you fill up your car. As for establishing an effective Quantum Energy Field to protect against electromagnetic radiation, I learned that I would have to invest another $600 because "four disks need to be placed in the corners of the room in the shape of a quadrangle." Oy vey.

As far as Longevity Village goes, it seems that through natural selection inhabitants are blessed with genes that code for a protein, apolipoprotein E, that plays a role in a number of important biological processes linked with health. And Alpha Spin? All it can do is make your head spin with nonsense.

EMAIL WARNINGS

"This is no joke!" As soon as I see that phrase pop up in an email, I know what's coming. I'm going to be warned about some nasty substance that is unraveling the very fabric of society. Like margarine. It is "one molecule away from plastic," a widely circulating email proclaims. Even flies are smart enough to stay away from it. We also have to be on the lookout for moldy pancake mix, which apparently is lying in wait to kill us. Sodium benzoate, a common preservative, can trigger Parkinson's disease. And the MMR vaccine for children? Better forget it. Trading in mumps, measles, or rubella for autism is not an attractive proposition.

These warnings, often forwarded by good Samaritans looking out for our welfare, are generally based on some sort of misinterpretation of scientific research. But not always. Margarine being "one molecule away from plastic" is just plain nonsense. Plastics are composed of long molecules called polymers, while margarine is a blend of fats and water. There is no chemical similarity between the two. In any case, being "one molecule away" is a totally meaningless expression. Substances are made of molecules, which in turn are composed of atoms joined together is a specific pattern. I suppose one might say that hydrogen peroxide, H_2O_2, is one atom away from water, H_2O, but even this is meaningless. That extra oxygen atom changes the properties of the substance dramatically. Stick your finger into a bottle of pure hydrogen peroxide and you will quickly experience the effect of that extra oxygen.

Even if margarine had some chemical similarity to plastic, which it does not, its properties could still be dramatically different. Slight alterations in molecular structure can account for very significant changes in properties. As far as flies staying away from margarine goes, I have yet to see a study confirming the allegation. In any case, our dietary decisions should not be based on the dining habits of flies.

I must admit that I do prefer butter over margarine, but this has nothing to do with plastics or flies. Yes, I'm perfectly aware that butter has more of the "bad" fats, but it also has more of the good taste. It comes down to a matter of quantity. If you are eating so much margarine or butter that the difference in saturated fat content makes a difference in the ratio of your total saturated-to-unsaturated fat intake, well, then you are eating too much of either one!

The warning about pancake mix is on a firmer scientific footing. The email refers to a newspaper advice column which described a fourteen-year-old boy's severe allergic reaction after eating pancakes made from a mix that had been in a pantry for a while. In all probability the reaction was to a mold that had contaminated the mix. Such reactions are rare but very real. The scientific literature does record a case of a young man who died after eating pancakes made from a mix that had been sitting open in a cupboard for two years. But the victim had a history of allergies, including to pets, molds, and penicillin. When the pancake mix was analyzed, a variety of molds including penicillium, fusarium, mucor, and aspergillus were found. For the vast majority of the population, mold in old pancake mix is not a life-threatening situation. But in general, it is not a good idea to eat food that has been sitting around for years. Make pancakes from fresh ingredients. Then your only worry is whether to top them with butter or margarine.

Sodium benzoate has been a controversial preservative ever since Dr. Harvey Wiley's "Poison Squad" sat down to dinner in 1902. Wiley, then chief of the Bureau of Chemistry, a forerunner to the FDA, enlisted volunteers to dine on meals laden with some of the food additives in common use at the time. He became alarmed when large doses of sodium benzoate caused adverse effects, but Congress refused to ban the additive after follow-up studies using more realistic amounts of the chemical failed to reproduce Wiley's results. The controversy over sodium benzoate reignited in 2018 when Professor Peter Piper at the University of Sheffield tested the effect of benzoate on yeast cells and discovered that the preservative damaged DNA molecules in the mitochondria, the cells' energy-producing machinery. This kind of damage has also been seen in Parkinson's disease patients, but suggestions that the small amounts of benzoate used to preserve certain foods can cause the disease are way off-base. And let's remember that benzoate isn't added to food for the fun of it. It can prevent molds from growing! And as we have seen, these can be nasty organisms.

Molds can be nasty alright, but not as nasty as the viruses that cause measles, mumps, or rubella. Advising people to shun vaccines that protect against these viruses is a far more serious business than scaring them about margarine, pancakes, or benzoates. But back in 1998, the warning appeared to have some substance to it. Dr. Andrew Wakefield, along with twelve colleagues, published a paper in *The Lancet*, one of the world's premier medical journals, suggesting a possible connection between autism and the MMR vaccine. The paper described a dozen cases in which children had supposedly developed autistic symptoms shortly after receiving an MMR vaccine. Controversy erupted almost immediately, with many parents refusing to allow their children to be vaccinated. But when the *Times* in London launched an investigation into the affair, a frightening picture emerged. It turned out that some of the children's parents had been recruited for the study by an attorney who was preparing a lawsuit against the manufacturer of the vaccine, and that Wakefield had been personally paid handsomely by the Legal Services Commission that was also funding the potential lawsuit against MMR manufacturers.

The *Times* managed to unearth a number of other irregularities associated with the *Lancet* paper, which was eventually retracted by the journal. Dr. Wakefield was charged with professional misconduct by the General Medical Council in Britain and lost his license to practice medicine. He now lives in the U.S., where he has been warmly welcomed by the anti-vaccine movement. Anyone deciding against MMR vaccination based on the flawed Wakefield study is making a mistake. Coming down with measles, mumps, or rubella is certainly no joke!

HEALTH FOOD STORE FOLLIES

"My mother hasn't been dieting but has lost about twenty pounds in a month and she feels tired all the time. What can you recommend?" That's the question my students were to ask when approached by a salesperson in a health food store. It was all part of an assignment designed to investigate the reliability of advice provided by such establishments. Of course, there is only one reasonable answer to such a question and that is "tell your mother she should go and see a physician." And that is the answer that was given in about 50 percent of the cases. But the rest of the salespeople offered up a bewildering array of vitamins, "glandulars," protein powders, creatine supplements, chromium pills, exotic juices, magnetic bracelets, and drops of "aerobic oxygen." One advised switching immediately to drinking distilled water and eating only organic produce. Another came up with a truly startling diagnosis for the weight loss: "I bet your mother has a microwave oven in her kitchen and uses Teflon pans."

Although this was to be only a learning experience for students, the results were so interesting that I thought they merited publication. Unfortunately, though, since we had not planned on a rigorous study the students had not documented the "evidence" to a degree that would pass scientific muster, and we had not considered what an ethics committee might say about our project. So you can imagine my interest when I saw a recent paper in a peer-reviewed publication, *Breast Cancer Research*, with the title: "Health food store recommendations: Implications for breast cancer patients." Edward Mills and his colleagues had sent research associates of varying ages into health food stores in the Toronto area to browse the shelves until approached by an employee. They would then reveal that their mother had been diagnosed with breast cancer and proceed to ask for suggestions about what to do. The study had been approved by the institution's ethics committee and the research associates had been schooled in what questions to ask and what information to divulge. Each encounter was immediately documented after the researcher left the store.

Altogether thirty-four stores were visited, and in twenty-seven of them recommendations were made for the use of some sort of "natural health product." A total of thirty-three different treatments were suggested. These included the usual vitamins, a cacophony of botanicals, shark cartilage, garlic, grape seed extract, dehydrated vegetables, antioxidants, herbal teas, and, in one case, a preparation made from the Venus flytrap. Only one common feature links all these products: a lack of scientific efficacy for the treatment of breast cancer. The most commonly recommended product was Essiac, an herbal remedy that contains burdock, Indian rhubarb, sorrel, and slippery elm. It was first popularized in the 1930s by Rene Caisse, a Canadian nurse (Essiac is her name spelled backwards) who claimed to have learned of this cancer cure from Indigenous medicine men. Essiac has actually been tested both in humans and animals and has shown no anti-tumor activity.

Fewer than a quarter of the employees discussed the potential for any interaction between prescription drugs and natural health products, and only three employees mentioned the possibility of any adverse effects. Two employees claimed that their recommended products could cure breast cancer, and perhaps most disturbing was the suggestion by one salesperson that tamoxifen, an established breast cancer treatment, be discontinued. The recommendations did not come cheap; on average they would set the patient back $58 a month, with some costing as much as $600. Finally, only 44 percent of the employees recommended visiting a health care professional and, even then, the majority suggested a visit to a naturopath.

By now, some of you may be thinking that the study in question was carried out by some medical or pharmaceutical establishment types, bent on showing the frailties of the natural health movement. As the argument often goes, these people are worried that their "slash, burn, and poison" methods used to treat cancer are going to be replaced by gentle, effective natural methods. Profits are at stake, so they supposedly look to design studies that will discredit the "opposition." Well, this study was not funded by or carried out by physicians or drug companies. Edward Mills, the lead author, was at the time director of research at the Canadian College of Naturopathic Medicine! He, like many others involved in the pursuit of effective natural products, was disturbed by the unreliable and sometimes dangerous advice given in health food stores. Spreading nonsense certainly does not further the cause of complementary and alternative medicine.

The ideal way to cut down on the confusion that surrounds the treatment of breast cancer is to reduce the need for any treatment. In other words, prevent the disease in the first place. Presently, there is no consensus on the relationship between diet and breast cancer, but being overweight is an established risk factor. This is probably because fatty tissues are not only capable of storing estrogen but are also adept at converting male hormones produced by the ovaries and adrenal glands into estrogen. High levels of this female hormone have been conclusively linked to some breast cancers.

Estrogen production can also be reduced with physical activity. Post-menopausal women who exercise moderately for roughly two hours a week can reduce their breast cancer risk by as much as 20 percent. Younger women who work out for at least four hours a week during their reproductive years can reduce it by 50 percent! Exercise works for young girls as well. We know that early onset of menstruation is linked to more intense hormone exposure throughout life and therefore to breast cancer. A large study of elementary school girls found that just five hours of exercise a week can delay puberty and thereby lower the threat of the disease. Exercise is good for you. And that advice is based on science. It is a lot more reliable than what you may hear in a health food store. Just ask the former director of research at the Canadian College of Naturopathic Medicine.

JILLY JUICE

I don't get angry easily. Yes, I get annoyed when I see homeopaths promoting absurd "nosodes" as alternatives to vaccines, chiropractors "adjusting" babies' spines, televangelists hawking Miracle Spring Water for protection from disease, and scientifically illiterate bloggers claiming that any food ingredient with an unpronounceable name should not be consumed. My annoyance, however, switched to anger when I became aware of two particular cases in which farcical "health advice" was being offered by self-proclaimed experts.

Brittany Auerbach, whose social media moniker is Montreal Healthy Girl, presents herself as a naturopathic doctor. This is based on having taken an online course from the unaccredited Institut de formation naturopathique du Québec (IFN), whose president claims to have a degree from the nonexistent Lincoln College of Naturopathic Physicians and Surgeons. Students who take the IFN's online course never see a patient and cannot be licensed as naturopaths. They are of course still free to produce videos and pontificate on all sorts of subjects in which they have no expertise, as Auerbach routinely does.

The video that really ticked me off was one in which she claimed that "cancer is actually a good thing," because it is a warning that the body is too acidic, a condition that will result in death unless an alkaline diet is followed. She maintains that "all diseases are reversible with the proper lifestyle changes and positive outlook," that HIV and Ebola infections can be treated with colloidal silver, and refers to chemotherapy and radiation as "hideously ludicrous."

Physicians, Auerbach argues, induce panic in cancer patients because of their ignorance and lack of understanding about cancer. She, on the other hand, claims to be "a certified naturopathic doctor who has done extensive research on cancer" and knows that "making your body ideal for healthy cells will cause all cancer cells to die over time." This drivel is found in a letter to her own uncle after he had been diagnosed with throat cancer. Instead of chemotherapy, which according to an oncologist had a chance of beating the disease, the man chose to treat himself with wheatgrass juice and an alkaline diet. It didn't work. One wonders how many others have been seduced by such apparent simple solutions to a complex problem.

Brittany Auerbach may be scientifically confused, but she still takes a back seat to Jillian Mai Thi Epperly, who soared to new heights with her stunning ignorance and outlandish claims. I had not heard of this spectacularly wacky woman until I was alerted to her appearance on *Dr. Phil.* Never before have I heard such concentrated hogwash in such a short time. Epperly claimed that the cure for every disease from AIDS to cancer is over-salted fermented cabbage juice that rids the body of all disease-causing toxins and parasites by triggering "waterfall" diarrhea. Never mind that the high salt concentration can trigger dangerously high blood pressure. That's not all. Drinking Jilly Juice will allow us to expand our life expectancy to 400 years! It even "cures" homosexuality and, get this, regrows missing limbs! Epperly proudly admits she has no relevant education, which according to her is a plus because it is educated doctors who have failed to curb all the diseases that afflict us.

I think there are two possibilities here. Epperly is either missing a few screws, or she came up with a devilish scheme to get attention that translates to monetary gains by making claims that are more outlandish than the most absurd lunacies that permeate the Internet. Indeed, Jillian managed to slither onto the set of *Dr. Phil* and undoubtedly procured some new paid subscriptions to her website, where she rambles on about her miraculous juice. The host did appropriately discredit Epperly, who would need a fair bit of education to rise to the level of being uneducated, but still, giving her exposure will cause some desperate people to follow her insane advice.

Neither of these women is spreading benign poppycock. They both steer people away from effective treatments, spread false hope, and assail science. Auerbach says that if she were diagnosed with cancer she would not even think of conventional therapy. Epperly claims to have a protocol "to reverse 100% of all health issues" and advocates giving Jilly Juice to babies. I claim that both are full of baloney. Or bunkum. Or balderdash. Take your choice. I have other words too.

There is some good news, though. At the McGill Office for Science and Society we have managed to bring the antics of Montreal Healthy Girl to the attention of the media, triggering television and radio investigations that resulted in her taking down the asinine cancer videos. I am also pleased to see that there are a number of groups in the U.S. that are trying to stop Epperly from spreading her mental muck on social media, and that the Federal Trade Commission sent her a letter asking for proof of her claims, which prompted Jillian to remove them from her Facebook page. Unfortunately, it is hard to take legal action because she is not selling the juice. She is only selling claptrap, which unfortunately is legal. Perhaps now you understand my anger.

CANCER AND CARNY TRICKS

"Up your nose with a surgical clamp!" That was João Teixeira de Faria's prescription for treating breast cancer. I first came across this Brazilian "healer" known as John of God in 2005, when he was featured on the ABC television program *Primetime Live*. John, who has all of two years of schooling, claimed that he was only an instrument in God's divine hands and that during a healing session his body is taken over by the spirits of long-dead physicians who guide his actions. Judging by the instructions they provide, it seems these physicians missed quite a few classes in med school. John, however, did not solely rely on departed physicians for advice. King Solomon can also be called upon when needed. The spiritual connections also allow John to diagnose a patient with just a glance.

Once the diagnosis was made, the healing procedure began. It some cases it was "visible," in others "invisible" spiritual surgery. If the patient chose invisible, they were directed to a room to meditate while the spirits did their work. "Visible surgery" involved sticking a surgical clamp up the patient's nose. This looks very impressive but is nothing but an old carny trick, usually performed with a long nail and a hammer. Any anatomical text will reveal that there is a roughly four-inch-long passage up through the nasal cavity that is quite ready to accommodate a foreign object without any harm.

I recently saw an entertaining performance of this effect in front of the Ripley's Believe It or Not museum in New York, of course without any implication of therapeutic value. The lady standing beside me gasped and exclaimed, "I don't believe it," despite just having examined the nail and having witnessed the show in front of her nose. It is easy to see how desperate people can be led by the nose to believe some sort of supernatural power must be involved and that someone capable of carrying out such a feat can perform other miracles as well.

John maintained that the success of his treatment hinged on the patient abstaining from drinking alcohol, eating pork, and having sex for forty days after treatment. That was a convenient "out" in case no miracle occurred. Patients were also able to be healed even if they were unable to travel to Brazil. All that was needed was a surrogate willing to undergo the spiritual surgery. No evidence for this remote healing was ever provided.

The forceps up the nose was not the only trick up John's sleeve. To treat nervous conditions, he appeared to scrape the patient's eyeball with a knife, while other problems were doctored with small random cuts on the body. As the *Primetime* cameras recorded, none of the patients showed any sign of distress after these rather invasive procedures. Quite the opposite. They believed they had been helped. Belief is a powerful tool indeed! There is a long history of television faith healers having the infirm throw away crutches and walk away unaided. Of course no cameras are present when they crumple to the floor backstage. An adrenaline rush stimulated by faith can produce amazing effects.

In an attempt to provide a critical view of John's antics, the producers invited two experts, cardiac surgeon Dr. Mehmet Oz and James Randi, the world's leading investigator of "paranormal" phenomena. Oz was likely chosen because he was a proponent of various alternative therapies such as therapeutic touch and reflexology and would likely be somewhat sympathetic to faith healing, perhaps adding an air of legitimacy. Randi was invited as the token skeptic.

Dr. Oz appeared repeatedly in the hour-long show, basically echoing the refrain that science doesn't have all the answers and that other forms of healing need consideration. Science of course doesn't claim to have all the answers, but it does look for evidence before jumping on a bandwagon. Randi, who could have provided evidence for methods of trickery and for psychological manipulation, was given a total of nineteen seconds on the show after being interviewed for hours. Why? Because the possibility that cancer can be healed by penetrating the nose with surgical forceps by a healer chosen by God makes for better television than declaring him to be a self-delusional simpleton or a calculating fraud artist.

In any case, it is a fact that people spent thousands of dollars to travel to Brazil to be poked, prodded, and scraped. Why? Because desperate people do desperate things. And many will provide alluring accounts of benefits. As Benjamin Franklin said, "There are no greater liars than quacks — except for their patients." Nobody wants to admit that they were swindled by some peasant who put tweezers up their nose. It is more comforting to believe that they were helped.

But what about the ones who gave up conventional care to go this route because they believe it to be more effective? Like South African singer Lisa Melman, who refused breast cancer surgery to be treated by John in 2005 and appeared on Oprah Winfrey's show singing his praises. Unfortunately, in 2012 she stopped singing forever, succumbing to the disease of which she was supposedly cured.

In an ironic twist, in 2015 John of God complained of a pain in the stomach to his cardiologist. Yes, the medium who claims wondrous healing powers had a cardiologist who without fanfare years earlier had implanted three stents in John's narrowed arteries. After the complaints of abdominal pain, he sent his patient for an endoscopy that revealed a tumor. A ten-hour surgery, not the spiritual variety, was followed by extensive chemotherapy. John had no problem affording the treatment. He accrued a fortune from donations and sales of blessed water and magic triangles.

When asked why he did not heal himself the way he is able to heal others, he replied with the stunning rhetorical question, "What barber cuts his own hair?"

John of God's bizarre story does not end with his fraudulent claims of curing disease. It turns out that through his fabled career this apostle of God was decidedly ungodly. In 2018, he was accused by a number of women of sexual abuse and within a short time the number of accusers swelled to over six hundred. This included his daughter, who called him a "monster" and alleged that she was beaten and raped by her father until she ran away from home when she was fourteen years old. In 2019, Faria was sentenced to nineteen years, essentially life imprisonment, for rape. A year later, due to poor health and the pandemic, he was transferred to house arrest. If there is a Heaven, John of God will not gain entry.

TREPANATION

The camera closes in on the electric drill boring a hole through the skull. Blood gushes from the wound and splatters everywhere. The latest horror movie? No. This scene is from a 1970 "documentary" called *Heartbeat in the Brain*, a true cinematic classic. The message of the film is clear: if you want to achieve a higher degree of consciousness, you need a hole in the head!

In this case the hole belongs to Amanda Feilding, a British artist who calmly shaves her head, makes an incision with a scalpel, and then attacks her skull with an electric drill. Joey Mellen, the cameraman, is a veteran of self-administered surgery and already sports a hole in his head. Why did these two decide to ventilate their brains in this fashion? Because they bought into a truly remarkable theory advanced by Bart Huges, a Dutch medical student who failed his finals and never managed to get his degree. Perhaps, given his ideas about brain function, we shouldn't be too surprised at Huges's lack of academic success. Plainly stated, according to this Dutch "savant," the enemy of clear thinking is gravity! Much of man's misfortune, he claims, can be traced to the time when the "ape first stood upright." We'll ignore for now the fact that while we may have evolved from a common ancestor, we did not, except maybe for Huges, descend from apes. In any case, Huges contends that when man became upright, there was a loss of blood from the brain because the heart now had to propel blood against the force of gravity. This had severe consequences. The brain had to take emergency measures to ensure that the parts essential for survival received sufficient blood. Capillary vessels leading to less important areas of the brain were constricted, reducing blood flow, and as a consequence, the delivery of glucose and other nutrients. Thinking, it seems, is not one of the brain's more important functions. Luckily, though, there is a remedy for brain activity that has been impaired by gravity!

According to Huges, if a proper level of consciousness is to be reached, blood volume in the capillaries has to be increased. The most obvious way to do this is to stand on one's head and allow gravity to undo the mischief that it caused in the first place. Bats, by this reasoning, should be among the most intelligent of creatures. Jumping from a hot bath into cold water also does the trick. So do certain drugs, like LSD. This substance, Huges says (without any evidence), constricts the veins in the neck, impeding the exit of blood from the brain and thereby makes us smarter. But these are temporary measures. A permanent higher level of consciousness can be achieved by boring a hole in the skull, technically known as trepanation. The convoluted theory explains that if the brain in not constricted by the skull, the heart can more easily pump blood into it. Huges refers to babies' skulls, which are not yet closed and where the "pulsation" of the blood in the brain with each heartbeat can be clearly seen. Not clearly seen, though, is the relevance of this argument. Neither is it apparent why people who "unseal" their skulls for "psychic buoyancy" feel that their rationale for doing this is buttressed by the discovery of trepanned skulls around the world.

It is certainly true that skulls with holes in them have been dug up by archaeologists. Some of these date back to the Stone Age. But this hardly means that our ancestors fortuitously discovered the secret to enlightenment. It does mean that some primitive people did have surprising surgical techniques when it came to treating skull fractures caused by slingshots or clubs. Bone fragments were neatly excised to prevent brain damage. Careful examination of the bone structure around these wounds shows that in some cases the bone healed and the patient survived! Perhaps in some cases such surgery was also performed as a treatment of mental illness by means of releasing "evil spirits."

Can there be anything to the suggestion that a hole in the head is good for us? Brain function is actually related to blood flow, not blood volume. And drilling a hole in the skull does not increase blood flow. Many patients, of course, have had holes drilled in their skulls for various types of brain surgery but none of them report any sort of enhanced mental clarity. A typical example would be temporary removal of part of the skull after aneurysm surgery to accommodate swelling of the brain. In one fascinating Australian case, the surgeon misplaced the bone that was removed and eventually the hole was covered with a titanium plate. The patient sued the hospital because she said she couldn't get over the feeling that part of her skull had been fed to a dog. Maybe if they had left the hole uncovered she would have reasoned with more mental clarity.

And just what kind of enhanced wisdom did Amanda Feilding achieve? Hard to say. Perhaps not a great deal if we judge by the fact that in 2000 she felt the need to repeat the trepanation. But maybe she had gotten a boost in intellect from the original hole because this time she found a neurosurgeon in Mexico City to drill her skull. One has to wonder if he too had a hole in his head. Amanda also ran for Parliament twice (a hole in the head of course is not a disqualifier in this pursuit) on a platform that trepanation should be covered under the National Health Plan. She garnered 49 votes the first time and 139 the next. Obviously there are a number of voters in Britain who should have their heads examined. And some on this side of the pond as well.

Back in the 1970s, Peter Halvorson, a trepanned American, founded an advocacy group to promote holes in the head. He claimed that in about 10 percent of the population the intracranial seams that we all have as children do not heal and the natural openings provide for enhanced brain function. John Lennon, he says, was an example, as are physicians! It is their positive pulse pressure that allows them to get through the extensive education required to become a doctor. An education that I imagine prevents them from mangling their skulls. In any case, people who do not have these natural openings in the skull can still make themselves smarter through trepanation. I think I will stick to drilling holes through pseudoscientific ideas.

BORBA'S NONSENSE

I think in recent years "nonsense" may have become the most frequently used word in my vocabulary. I can't even begin to guess how often it rolls off my tongue in response to questions I'm asked on my radio show or after public lectures or via email or telephone. Can drinking "oxygenated" beverages increase stamina? Is it true that lemon juice can flush toxins out of the body? Can you cure disease by tuning into its vibrational frequency? What else can you say, but "nonsense"? (Well . . . maybe there is another word.) But I must admit I was a little stymied when asked about the claim that Borba Gummi Boosters can help "skin clarity."

Now, the only gummies I know are of the bear or worm variety, made basically of cornstarch, corn syrup, sugar, and gelatin. Throw in some food dyes and some citric acid for sour flavor, and you've got a child's dream and a parent's nightmare. But what is a Borba Gummi? As it turns out, Borba is not a what, but a who. Actually, more than just a who. It was a whole company named after the who. And who is the who? That would be the remarkable Scott-Vincent Borba.

After obtaining a B.Sc. in psychology, Borba became a licensed aesthetician and founded a company to market a line of products that purport to improve skin quality both from the outside and the inside. The company produced drinks, gummy bears, and jelly beans that "conspire with your bloodstream to nurture skin where it starts." And of course, these went hand in hand with creams based on Borba's Fiber-Knit Technology, which was said to incorporate "nutrient-infused micro-hydration and natural hydrolyzed spandex fibers that envelop the epidermis from above and below to create your optimum skin condition." Needless to say, the cream also claimed to "revitalize epidermal firming memory." Of course, as Borba told us, the cream works best when used in conjunction with his Skin Balance Water.

There is no doubt the man is a marketing genius. While he may not know much about chemical formulas, he sure knew the formula to sell products: take some aging baby boomers who want to retain their youth at virtually any cost, fling some scientific terms like "reverse osmosis," "antioxidants," "calorie-free," and "micro-hydration" at them, and offer up some product that contains vitamins, minerals, and a smidgen of some natural ingredient that has been in the news. Pomegranate, acai, and grape seed work well, but botanicals that your market hasn't heard of are even better. Cherimoya and longan fruit are great, and if you can toss in a few words about how these have been used for millennia by the Chinese, even better. And then claim that you have found the fountain of youth, slap on a hefty price, and watch your sales zoom.

How did Borba get into this game? Well, let's let him tell the story. "Honestly, I was rushing out of my house to catch a flight and I tripped with a handful of supplements, a liter bottle of water, and a couple of skin care products in hand. They all fell into a puddle together and that was my epiphany moment." Just why he was rushing around holding supplements and skin care products isn't clear, but what is clear is that Borba went on to sell Skin Balance Waters and gummy bears spiked with a blend of antioxidants, vitamins, and plant extracts that claim to restore lost beauty. Of course, you were told to choose the right product, depending on your needs. A different fortified water would be required depending on whether you needed clarifying, age defying, replenishing, firming, or skin calming effects. I suppose if you had more than one of these problems, there was a risk. You could become waterlogged. And there was another effect. A thinner wallet. The drinks were expensive.

Was there a chance that the drinks and gummy bears could really do what they claimed? Not much. No government approval is needed to market such products because they come under the aegis of dietary supplements rather than drugs. That is a strange distinction, because it seems to me that if you make a claim of improving the elasticity of the dermis and removing epidermal toxins, you are making a drug claim. And Borba's Gummi Boosters did claim to "help skin regenerate its natural support system, as they removed toxins and improved skin clarity with cultivated bio-vitamin complex." This special complex was nothing other than a run-of-the-mill mix of vitamins and inconsequential amounts of grape seed, green tea, chamomile, and acai extract. A 375-gram bag of these wonder gummies was priced at $25, and customers were urged not to exceed a bag a day.

Note the use of the past tense in the preceding. That is because Borba's dietary supplements have disappeared from the marketplace. Why? I suspect if they had delivered the goods, they would still be around. But Borba hasn't disappeared. He sells books such as *Cooking Your Way to Gorgeous: Skin-Friendly Superfoods, Age-Reversing Recipes, and Fabulous Homemade Facials*. I think my frequently used word still applies. Nonsense.

THE GERSON FOLLY

What sort of treatment do you think cancer patients would receive at the Gerson Institute in San Diego? Actually, they would receive no treatment at all, because the Gerson therapy is not sanctioned in the United States. But they would receive plenty of information about traveling to Gerson clinics in Mexico or Hungary, as well as about providing basic Gerson care for themselves at home. The institute does not limit itself to providing information about cancer. It seems the Gerson therapy is effective against virtually every disease. How can this be? Because "it restores the body's incredible ability to heal itself with no damaging effects, and rather than treating only the symptoms of a particular disease, it treats the underlying cause of the disease." Right. And the tooth fairy leaves coins under the pillow.

Cancer is a terrible disease that often defies conventional treatment. But the failure of science-based medicine can mean success for the marketers of "alternative therapies" who are unencumbered by the need to furnish evidence. They just have to clamor about how conventional doctors slash (surgery), burn (radiation), and poison (chemotherapy) their patients, hastening their demise, while they offer kinder, gentler, lifesaving "natural" treatments. Desperate patients, they well know, will do desperate things. At any cost.

The Gerson Institute and Cancer Curing Society, as it officially calls itself, adorns its seductive brochure with the credo "healing with nature." Aside from the absurd but appealing notion that "nature" is more adept at healing disease (which it incidentally causes with reckless abandon through natural bacteria, viruses, fungi, and molds) than research-based medicine, one has to question the "natural" aspect of the Gerson regimen.

Is the squirting of coffee up one's rear end "natural"? What about gulping desiccated liver capsules? Or administering ozone rectally? All these have been part of the program. To say nothing of drinking several glasses of raw calf liver extract a day! That lunacy was given up after several patients' deaths were linked to a bacterial infection associated with the extracts. The foul liver juice was replaced by a more taste-bud-friendly green leaf–apple juice blend, a dozen glasses of which have to be downed to "flush the toxins" responsible for cancer out of the system. Just what these toxins are is never addressed. But to make sure they are eliminated, patients are also dosed with pancreatic enzymes, iodine, vitamin B12, niacin, thyroid hormone, potassium, coenzyme Q10, and organic flaxseed oil. Of course all of these bizarre interventions would be acceptable if the treatment worked. Let's face it, conventional chemotherapy is no picnic. But there is a difference. Chemotherapy at least has a chance of working.

As the name suggests, there actually is a person behind the Gerson therapy. An established physician, Dr. Max Gerson fled his native

Germany in 1936 when the Nazis came to power and eventually settled in New York. As a young doctor he had been tormented by migraines and had sought relief by experimenting with different diets. He traded in his wursts, schnitzels, and sauerbraten for a plant-based diet that apparently resolved his migraines. Gerson theorized that contamination with artificial fertilizers and pesticides was responsible for his misery. He began to prescribe his "natural" plant-based diet to other migraine sufferers, who soon claimed to experience all sorts of additional benefits, including resolution of tuberculosis. Needless to say, there was no objective evidence that any patients had actually been cured in this fashion. How could there be? TB, a bacterial infection, cannot be cured by diet.

And then Gerson had an epiphany. If TB responded to his regimen, why not cancer? By 1958 he had published his book, *A Cancer Therapy*, in which he described curing fifty patients of terminal cancer. That astounding claim prompted the U.S. National Cancer Institute to undertake a review of Gerson's cases, reaching the conclusion that the validity of the cancer diagnoses and the supposed cures could not be substantiated. Gerson retorted that the review had been unfairly influenced by the "cancer establishment," for the simple reason that his natural cure was a threat to the grotesque profits realized by the pharmaceutical industry from its expensive but useless chemotherapeutic drugs. That tired old refrain has practically become the anthem of the "alternative medicine" community.

The problem with the Gerson therapy, as now promoted by his daughter Charlotte and practiced in the Mexican and Hungarian clinics, is not that it is scientifically implausible, nor that it is tortuous to follow, nor that it is repugnantly expensive. The problem is that there is no evidence that it works! The Gerson clinics make all sorts of claims about euphoric patients returning home, cured of their disease. But no follow-up is ever carried out. And whenever independent researchers have tracked Gerson patients, they have found that most had succumbed to cancer within five years of having been "cured" of the disease.

There is even less information available about the success or failure of the home version of the Gerson therapy. Administering coffee enemas at home may be a bit of a challenge, but the juicing can be done. Not with any old juicer, though! No sirree. We are told that "Dr. Gerson's research indicates that it is imperative for cancer patients to have a two-step juicer with a separate grinder and hydraulic press. One-step juicers generally do not produce the same quality of enzyme, mineral or micronutrient content." Really? I don't seem to be able to find that bit of research in the peer-reviewed literature.

The Gerson website actually recommends a specific juicer that will run you in the neighborhood of $2,000. Surely, though, that's a bargain if it will help you beat cancer. Don't even think about buying a cheaper juicer, because as the Gerson Institute's captivating brochure tells us, "in fact some patients have failed to experience results simply by using the wrong juicer." Yup — that must be why they failed to cure their cancer. Wrong juicer! Those cutting-edge researchers at the Gerson Institute surely would not lie to us, would they?

MIRACLE MINERAL SOLUTION IS A NIGHTMARE

Malaria, AIDS, hepatitis, herpes, cancer. Terrible diseases. That's why thousands and thousands of scientists around the world, armed with advanced degrees, are engaged in research projects aimed at finding cures.

Now, ask yourself this question: what is the chance of a gold prospector, with no training in the health sciences, tackling a problem and finding an answer that has eluded the world's most renowned researchers? Furthermore, it's simple to administer, readily available and, to boot, also clears up acne, eliminates heavy metals, cures the common cold, and destroys the SARS-CoV-2 virus. I can tell you what probability I would attach to this miracle solution performing as claimed. Let me see now, how does "zero" sound?

There's nothing subtle about the name of this purported wonder: Miracle Mineral Solution (MMS)! Well, there are no miracles to be had. Or minerals. Admittedly, however, there is a solution. Not a solution to any problem, but a solution in the sense of a substance being dissolved in water. And that substance is sodium chlorite, a common disinfectant and bleaching agent. Its chief promoter, Jim Humble, is either a brilliant inventor, a self-delusional scientifically bewildered simpleton, or a cunning scoundrel. Take your pick. I know which box I would tick.

In a decidedly non-humble fashion, Humble claims that "this breakthrough can save your life, or the life of a loved one." He then brags that his discovery is the answer to AIDS, hepatitis A, B, and C, malaria, herpes, TB, most cancers, and many more of mankind's worst diseases. Of course you may not have heard of this revolutionary treatment because it is being hidden from the public by those devilish pharmaceutical companies whose profits would be destroyed if the word got out about all diseases being cured in such a simple fashion.

Let's just trace how this visionary, this wonder-worker, this mental colossus, discovered the gift "that would shift the course of human health history forever." Incidentally, MMS wasn't this amazing philanthropist's first gift to humanity. That was the automatic garage door opener, which humble Humble supposedly invented, although I can't find any documented evidence for this claim.

In any case, the MMS saga begins in the South American jungle, where our hero was prospecting for gold when two of his men fell ill with malaria. With no prospects for immediate medical help, Humble had to resort to his razor-sharp wits. Actually, I don't think there was much chance of any cuts being inflicted. The sodium chlorite solution he had brought along to disinfect water obviously killed bacteria, our champion thought; maybe it would also destroy whatever was causing the malaria. So, he gave the men some of the solution and was stunned to see their symptoms vanish in just four hours. I bet he was!

Now Humble had a new calling, to rid the world of malaria. He started to treat sick South Americans but found that the sodium chlorite solution was only effective 70 percent of the time. Not good enough for this dazzling mind! He began to experiment with his concoction and discovered that when mixed with citric acid, the chlorite would be converted to chlorine dioxide, which turned out to be a superior treatment. Wow! Before long, Humble claimed to have registered 75,000 successful treatments of malaria with his miracle product.

Strange, but I can't find any of these spectacular results documented in the scientific literature. Wouldn't you think that the discovery of such a simple cure for malaria would merit publication? Surely a Nobel Prize would be in the offing! Ah, I know. It must be those dastardly jealous scientists, or the evil pharmaceutical companies that are preventing publication. Yup. Must be. As a supporter explains, "Humble had become so famous that two drug companies contacted the Minister of Health [in an unnamed country] and threatened to quit shipping drugs to the local hospitals if she didn't do something about the person claiming to be able to cure malaria." Those fiendish companies! It's a wonder they have allowed Humble to live. Actually, maybe they haven't. Attempts to contact him repeatedly fail. I'm told he is traveling the world, busily helping people. Helping them lighten their wallets, I suspect. If you want to know the details of his discovery, that is "how to manufacture it in your own kitchen, how to use it intravenously, how to cure colds in an hour, how to cure the worst of flu in twelve hours, how to treat cancer, AIDS and hundreds of other problems," you have to buy his book.

I'm not sure how to describe that epic work, but "comedic" comes to mind. Discussions about how chlorine dioxide "elongates the electron shell" of pathogens, and how its safety is confirmed by the fact that its "oxidation strength" of 0.95 volts is less than oxygen's 1.30 volts, amount to no more than mindless chemical chatter. I don't buy it. More importantly, Health Canada and the U.S. Food and Drug Administration don't buy it. And both urge consumers not to buy any version of Miracle

Mineral Supplement. Not only is there no evidence of efficacy for any condition, there is evidence of possible harm. Nausea, vomiting, and a life-threatening drop in blood pressure have been reported. Humble actually maintains that nausea is a good thing because it means the body is eliminating toxins, but if it's bothersome, he suggests it can be controlled "by eating cold apple slices that will absorb stomach toxins that have been dumped there." Like I said, comedic. But what is decidedly not comedic is the advice on some MMS websites that AIDS patients give up their drugs and resort to intravenous MMS.

Jim Humble went out looking for gold and it seems that at least figuratively he has found it. But it is fool's gold. MMS is not based on any reasonable science, has not been tested in any sort of randomized trials, and amounts to no more than a scheme to capitalize on the gullibility of the scientifically challenged and the desperate. Promoting the sale of this product is criminal.

Humble has also climbed aboard the coronavirus bandwagon. Here is his wisdom: "MMS (sodium chlorite activated with a food grade acid which then produces chlorine dioxide) kills most of the diseases of mankind. I don't know for sure about the coronavirus at this time — but we know that MMS kills viruses as well as pathogens of all kinds and is an immune system builder. There is much anecdotal evidence that says MMS has proven very effective in eradicating viruses including Ebola, Swine Flu, TB, and other respiratory diseases. Chlorine dioxide was completely effective against Anthrax in 2001, and used by U.S. Military for Ebola in 2014. It's been proven by the Red Cross in 2012 to eradicate malaria in just four hours, to name a few. I have been receiving feedback for over 22 years from people all around the world who have given testimony of how they recovered their health from a vast variety of diseases, many life-threatening, with MMS. Therefore, I have every reason to believe it can be effective in stopping and preventing the current novel coronavirus going around today."

Dr. Gabriel Cousens, a physician who is also a homeopath, has also gotten into the game. The man has a checkered history that includes loss of license for excessive prescribing, a lawsuit for the death of a patient caused by Cousens injecting "bovine adrenal fluid," and attempts to cure diabetes with a raw food diet. He believes that "a singlet oxygen atom of iodine" is the "most powerful antiviral substance available to us." This is a scientifically meaningless term, but nevertheless he sells Illumodine, which supposedly contains it. No idea what it actually is because the only available description is that it is organic (yup, organic iodine), radiation-free, 100 percent bioavailable, scalar-wave energized, algae-free, and 100 percent pure. What it is, is 100 percent nonsense. How someone with medical training can write such rubbish is puzzling. Cousens admits that he has no experience treating coronavirus infection; nevertheless he states, "I've found that 5–15 drops per hour [of his magical Illumodine] until the infection is gone is a reasonable protocol for most viral infections." No, this is not reasonable. There is no evidence that any form of iodine cures viral infections.

BRACE YOURSELF

This is a tale of two bracelets. One brandishes flagrant nonsense, the other flirts with some clever science. We begin with a perplexing question I was asked while wandering through a mall in Phoenix. "How would you like to experience the benefits of nature captured in holographic frequencies?" Sniffing that some delicious twaddle was coming my way, I answered that I was keen to resonate with nature.

It turned out all I had to do was put on a Power Balance bracelet "imprinted with frequencies that would interact in a positive way with my body's energy field." I would feel better, aches and pains would resolve, my balance would improve, and I would feel stronger. All because the unnatural vibrations produced by the likes of sugar, synthetic chemicals, and cell phones would be neutralized by the frequencies embedded in the wristband's hologram. Would I like proof? I was asked. Naturally!

I was then instructed to raise my right arm parallel to the ground and resist any attempt to push it down. I tried, but the salesman had no problem overcoming my resistance. He then slipped the bracelet on my left hand, and in spite of a convincing struggle on his part, my right arm hardly budged. "Energy is related to frequency," I was informed.

My protagonist, who was a rather muscular young man, was also sporting a Power Balance bracelet, which prompted me to ask how it was that its energy did not cancel out mine. This did seem to raise a point he had not previously considered, but he managed to mutter something about the benefits being greater if more unnatural frequencies had to be overcome. Do you eat only organic food? he asked. Not only, I answered somewhat ambiguously. His contented nod suggested the matter had been resolved.

Now it was my turn. I didn't think there was much point in discussing how it was indeed true that energy was proportional to frequency through Planck's constant, but that the frequency referred to was that of electromagnetic radiation and had nothing to do with the human body, which does not have any innate "resonance." Instead of trying to dam the river of the rapidly flowing pseudoscientific guck with scientific explanations, which I suspected would get us nowhere, I proposed my own experiment. I asked if the position of my left hand mattered, eliciting a chuckle. No, all that mattered was whether I was wearing the wristband or not. Good!

We would follow the same procedure as before, but this time I would put my left hand, which would either be sporting a bracelet or not, behind my back. His task was to determine if I was energized or not! Given our chat, he didn't have much choice but to agree. I suggested ten trials. He guessed right four times. Yes, "guessed" is the right term because there is no science here. But neither is there necessarily fraud. Perhaps in his eagerness to make a sale the young man didn't realize that he was subconsciously exerting less effort when I was wearing the bracelet.

How then do we explain the legions of athletes and celebrities who claim all sorts of benefits? Mind over matter is the real power in the

Power Balance Bracelet! As I subsequently learned, the marketers of the bracelet in Australia actually admitted as much after experiments, much like my ad hoc one, unmasked the product. Sales quickly went belly up. The bracelets are still sold there, but the claims are of the weasel variety: "Power Balance is a favorite among elite competitors, weekend warriors, and everyday fitness enthusiasts. The hologram is designed based on Eastern philosophies. Many Eastern philosophies contain ideas related to energy."

I'm more in favor of ideas related to science. And a new company, MyExposome, run by real scientists, has a good one. Supported by published proof of principle, the plan is to furnish people with a silicone bracelet that absorbs chemicals with which it comes in contact either from the air or from bodily secretions. Using gas chromatography–mass spectrometry, the bracelets will then be analyzed for some 1,400 chemicals, including controversial ones like flame retardants and phthalates. The company will not offer any advice on whether a particular chemical has any specific benefit or harm, because presently there isn't enough known to make such judgments. Hopefully, though, the data collected can eventually determine levels of exposure and any possible risks. MyExposome's scientific approach may give us real "power" to "balance" chemicals in our lives.

NO MAGIC IN QUACK CANCER TREATMENTS

Magic is the science of fooling people for purposes of entertainment, and magicians, be they professionals or amateurs like myself, take a great interest in the various methods that can be used to bamboozle people. Magicians, though, are honest charlatans and tend to get quite upset with dishonest charlatans who dupe people for purposes other than amusement. The most troubling forms of deception are the ones that deal with matters of health, particularly when it comes to preying on people's desperation. Cancer is a dreaded disease and there are plenty of charlatans ready to take advantage of its victims. Over the years I have often tried to alert the public to the various ruses with which these unsavory characters ply their trade. But, as is often the case, personal involvement takes investigation to a whole other level.

When my wife was diagnosed with glioblastoma multiforme, a terrible type of brain tumor, I did what most people do. I frantically searched the scientific literature for options and quickly discovered that there was little room for optimism. Inevitably, Googling brings up a host of "miracle cures," ranging from herbal remedies and electronic gizmos to coffee enemas. "Cutting edge," "breathtaking," "more powerful than any drug Western Medicine can offer," and "bombshell report" are phrases often encountered. There are stories of "stunned doctors who watched tumors disappear in just two weeks," and accounts of patients who failed to improve with "dangerous chemo and agonizing surgery" but experienced a miraculous recovery after opting for a "little-known natural serum in a tiny vial that has the power to crush the billion-dollar chemo and radiation industry." To find out what it is, you are urged to watch a video, and to do so quickly, "before the government, conspiring with Big Pharma, will force its removal." But after you've invested close to an hour, you learn that you have to purchase a book or some newsletter to have the secret revealed.

Another report describes a man whose "body remained riddled with tumours after eight brutal months of chemotherapy and had already bought a grave before every single tumour in his body was obliterated." It costs to find out how. Then there are "maverick" physicians who claim to answer to the Hippocratic oath, not drug companies, and "blow the lid off" Big Pharma's attempts to suppress a treatment "proven to be more effective than nineteen of their best selling drugs — but without side effects." There are numerous such websites featuring various "censored" cures, all claiming to have evidence that is being blocked from publication by drug companies trying to protect their turf. Right.

I received numerous emails from well-meaning people about treatments to explore, ranging from hemp oil and alkaline water to the

Amazing Amezcua Biodisc, which promises to cleanse chakras. One, Light-Induced Enhanced Selective Hyperthermia, seemed interesting enough to look into. What I found was not pretty.

Light-Induced Enhanced Selective Hyperthermia was actually a scheme cooked up by Antonella Carpenter, a septuagenarian alternative practitioner in Oklahoma who was not a physician but had some training in physics. She claimed to cure cancer by injecting a tumor with a saline solution of food coloring and walnut hull extract followed by heating the area with a laser. The treatment, she maintained, was 100 percent effective with no side effects. Of course, any claim of 100 percent efficacy is a hallmark of quackery, since no drug of any kind works in such a foolproof fashion. Even worse, Carpenter urged patients to stay away from oncologists and sometimes told them their cancer had been "killed," which was not the case.

As often happens, quacks unearth some legitimate process and then twist it out of proportion to hatch a money-making scheme. In this case, the legitimate process is photodynamic therapy. In general, the treatment of cancer involves some process by which cancer cells are destroyed while normal cells suffer less damage. Unfortunately, it isn't possible to avoid collateral damage completely, and cancer treatment via radiation or drugs is always burdened with side effects. In photodynamic therapy, the idea is to introduce a photosensitizer, a chemical that when activated by light interacts with oxygen to convert it into a very reactive form known as "singlet oxygen" that can destroy cells. The photosensitizer can be introduced intravenously and is followed by treating the tumor with long wavelength light via an optical fiber. Alternately, the photosensitizer can be injected directly into a tumor and then the area exposed to light. In either case, singlet oxygen is produced only within the tumor, minimizing damage to normal tissue. The process is applicable to certain types of tumors and is certainly not a cure-all for cancer.

It is this therapy that has been mangled by Antonella Carpenter, who according to investigators cheated cancer patients out of their money and gave them false hope. In spite of any evidence of her treatment having efficacy, supporters sprang to her side, claiming that her conviction on twenty-nine counts of fraud was carried out by a kangaroo court influenced by "the greedy and vindictive genocidal maggots who control the Cancer Industry and have the FDA and courts in their back pocket." They went on to say that "the medical mafia is hard at work twisting the truth and vilifying Dr. Carpenter and any other non-Allopathic practitioners of natural or alternative treatments as quacks." There's more. "Dr. Carpenter was vindictively targeted by the Medical Mafia and their Gestapo goons at the FDA for successfully curing dozens of cancer patients." No. She was targeted for subjecting cancer patients to a treatment that did not work and was claiming she had cured them. That is evil.

The truth is that there is no conspiracy to keep effective cancer treatments from the public. Such allegations are an insult to the thousands of researchers and physicians dedicated to solving the problem of this complex disease. As I well know, there is no real magic, only clever tricks to create the illusion that there is.

THE DETOX SCAM

I'm all patched up as I sit here writing this little piece. I've got a patch on my sole, one under my arm, and one on my derrière. Don't worry, I've not been injured in any way, I'm "detoxifying." Apparently just like thousands and thousands of Japanese and a growing number of North Americans. The patches, made by numerous companies, mostly in Japan, resemble large Band-Aids and claim to draw toxins out of the body. No reference is made to what sort of toxins are removed, but there is no shortage of claims about the results. Headaches, high blood pressure, kidney problems, arthritis, hair loss, fatigue, diabetes, heart disease, and giddiness are all helped. I can testify that at least in my case, the last claim is unfounded. When I read about how the patches supposedly work, I become more giddy.

These detox patches appear to be truly amazing devices. Not only can they draw poisons out of the body, they can also infuse various healing agents into the body. What sort of agents? Like those found in the Japanese loquat leaf, which we are told contains various vitamins, including "vitamin B17." Actually, there is no vitamin B17, but the term is commonly used to describe a discredited cancer "cure" also known as laetrile. The patch also has added vitamin C, which according to the label reduces cholesterol, blood pressure, and the risk of blood clot formation. No evidence for these claims exists, and even if it did, there would be better ways of introducing vitamin C into the body than through the bottom of the foot. There are also other gems in the formulation. Literally. There is powdered tourmaline, which "exerts a cleansing and liberating energy upon our entire nervous system, promoting a clearing and stabilizing effect." We are told that tourmaline is "one of the only minerals to emit far infrared heat and negative ions." Then there is amethyst, a "stone of psychic power" which "promotes tranquility and helps embrace your own intuitive wisdom." Doesn't seem to promote too much wisdom among the people who endorse this gobbledygook.

The main ingredient that is supposedly responsible for detoxication is "wood vinegar." This reddish-brown liquid is obtained by heating wood and condensing the vapors that form. It is a complex mixture of oils, tar, methanol, acetone, and acetic acid. Volatile components can be driven off by drying, leaving behind a gray powder, the "essence" of the detoxifying patch. This is the stuff that appears to magically draw toxins out of the body. And those unnamed toxins really do appear! At least in the pictures that accompany the product. The patch, which originally was white, becomes brown and sticky after being worn for a few hours. According to the literature provided, the brown sludge is formed by the poisons removed from the body. Nonsense. The stickiness is due to moisture combining with dextrin, a starch filler used in the patch. Remember mixing flour and water to make glue? That's just what is happening here. As far as the color goes, it is due to sweat reconstituting the wood vinegar.

So if foot patches are just so much poppycock, how is it that so many people feel better after the "toxins have been removed"? The same reason that people felt better after flocking to the elegant chambers of Mr. John St. John Long in Harley Street, London, in the early nineteenth century to be treated with a liniment made of turpentine, acetic acid, and egg yolk. Not unlike the foot patch ingredients. St. John Long had no medical training whatsoever, yet had supporters aplenty who were convinced that he had cured them of various ailments. The term "placebo" may have only been coined in 1920, but the effect of which Long made elaborate use has an extensive history. The ancient Egyptians, for example, alleviated abdominal pains by rubbing the belly with saffron powder and beer.

St. John Long could have had a long and fruitful career cashing in on the placebo effect, had he stuck to his ointments. But some cases required a more dramatic intervention. Internal disease, he proposed, could be treated by creating an external wound that would produce a discharge to carry off the malady. This is the philosophy he applied when a Mrs. Cashin, having heard of the wonderful cures effected by Mr. St. John Long, approached the healer, worried that her elder daughter would be afflicted by tuberculosis, a disease that had already claimed her younger sibling.

The quack proceeded to make an incision on the young lady's back to allow any incipient disease to escape. When a discharge, probably due to infection, was seen, Long expressed elation, declaring that "the wound was going on remarkably well, and that he would give a hundred guineas if he could produce similar favourable signs in some other of his patients." His elation did not last long as poor Mrs. Cashin soon expired. A coroner's inquest was summoned at which a number of witnesses spoke of the virtues of the accused's lotion for curing various complaints. Nevertheless, the jury found St. John Long guilty of manslaughter, but incredibly he got off with a fine that he immediately paid from a wad of bills in his pocket.

Within a month, he was back in court, this time accused of precipitating the death of the wife of a Royal Navy officer. Mr. Long was tried at the Old Bailey, but the jury found the evidence against him inconclusive. When he was found innocent, a great roar rose up from his supporters in the courtroom, who declared that Long's treatments had now been vindicated. The efficacy of the miraculous ointment, though, was called into question when three years later Long contracted tuberculosis and failed to cure himself of the disease. His former patients were unfazed by their hero's demise and collected funds for a memorial monument paying tribute to his talents.

So I suspect my saying that the detox patches amount to claptrap will not shake their devotees. But at least the patches are not dangerous claptrap. I've had no adverse effects from my little experiment. And for those of you interested in the technical details, it seems that armpits and soles are more toxic than bottoms.

YIKES! I'M INFESTED!

I have little insects inside me that are dining on my cartilage, bones, and muscles. It seems they invaded my body either from animals or from dirt. These bugs used to eat plants, I'm told, but because they've been genetically modified, they now eat us. I also have a type of worm in my blood vessels. These creatures come in couples with the female living in the male body. I also have an overabundance of vitamin C in my kidneys and an inflammation of the sciatic nerve caused by a plasma virus. My prostate gland is infected by a brown mushroom. My red blood cells are a little too big because of microbacilli that are either released by plants in my office or come from eating fruits that weren't washed properly. Apparently these bacteria like to eat the fat from the red blood cells, which then causes the blood cells to become bigger. I also have a viral infection in my right eye. And my muscles don't work properly because mushrooms have grown roots that tangle the muscle strings. I guess it's a wonder that I'm still alive.

I'm not too worried, though. The worms, bacteria, mushrooms, and viruses were not revealed by blood tests or CAT scans. They were diagnosed by a different kind of scan. I was informed of all the nasty action going on inside my body by a clairvoyant/naturopath who scanned me from top to bottom with her eyes closed, sensing, as she claimed, "life frequencies." Needless to say, my problems were "treatable."

Alright, let's rewind a little. This little adventure started with an email I received that intriguingly began with: "In the past those like me were called witch, saint, gifted, mutant, freak and more . . . but I have an extraordinary ability at being able to find elements and microscopic life such as bacteria, viruses, worms, parasites and algae in the human body, the earth's crust and so on." The writer assured me that this was not a hoax and was looking to be tested in exchange for a document attesting to her ability. I was game, and we discussed various ways that her abilities could be put to a test.

She told me that "when looking through a human, I see chlorine as yellow bubbles . . . radon as pale blue accumulation, copper as white." These claims really weren't testable, but we hit on something when she mentioned she could see germs in water and could distinguish between tap water, bottled water, and lake water. We settled on a challenge that involved randomly placing one of these waters into each of fifteen glasses. Her task was to identify the samples. She actually got eight correct, short of the ten that we had agreed would constitute a meaningful result. I asked if water that had no germs would be easier to identify and she thought that would be the case. So I set up four glasses that contained either tap or distilled water. She only got one of these right.

I thought we were now done with the experiment but was told that actually her main talent was diagnosing what was going on inside the body, and she was quite willing to demonstrate this ability. And so we began. Her very first words were "this is for entertainment purposes only," which was fine with me as I did think this would be quite entertaining. "There's a lot of carbon in your system, especially in the liver and the blood." Well, she got that right. All the proteins, fats, carbohydrates, and nucleic acids that make up our tissues are organic compounds, meaning their basic structure is built of carbon atoms. I don't think, however, that is what she had in mind.

Next I was told I have a lot of heavy metals in my lungs, like machinists who solder a lot. I think I soldered once in my life. I must have smoked in the past, she went on, because I have a lot of "old" carbon in my lungs. I have never smoked. She also diagnosed schistosomiasis, an infection with a parasite that causes my legs to be itchy. Schistosomiasis is an infection widely seen in Africa and Asia, never in North America. And my legs do not itch.

I have a purplish color in my liver. I was told that what you eat dyes your body and I must have been eating beets. Nope. Can't remember the last time I ate this vegetable. Then I was told that I often get pain in my rib area from coughing or from rotating movements but I should not worry because a chiropractor can easily fix that. I have no such pain, and should I encounter it, my choice of treatment would surely not be a chiro. After scanning me she did the same with two colleagues who were also filled with mites, insects, "phages," "microplasm infections," and who knows what else. In one case she even claimed to see a tumor, specifically in the left testicle.

While performing these scans she also revealed that she had the ability of communicating with the dead and volunteered to do readings for the three of us. It was amazing! Basically, because she got nothing right! My grandparents made no mention of the fact they died in the gas chamber, and my father must have been vigorously exercising on the other side because I was told he was a large, muscular man. Actually, he was shorter and smaller than me. I was also told that the reason I'm constantly searching for my keys is that they were being hidden by the mischievous spirit of a girlfriend I left for my wife. Nope and nope. No such girlfriend and I don't lose my keys.

Up to this point I had been sitting straight-faced without making any comment because I'm quite familiar with "cold reading," and the ability of "psychics" to capitalize on any reaction from their subject. But now I suggested we discuss the happenings and explained that she had been offtrack on virtually everything. At this point, she became agitated and asked why we had invited her if we were just going to waste her time, forgetting that she had sought the invitation. In any case, the clairvoyant then got up and muttered something about the failure being due to my skepticism that blocked her abilities, apparently not having foreseen this possibility. We never did get around to treatments, which I suspect were of the herbal variety. Next time, she said, she would seek out a microbiologist with an open mind and prove herself.

Anyone with a scientific background would of course recognize the garbled rhetoric we heard as total nonsense, albeit somewhat entertaining. But it was also clear that this clairvoyant/naturopath has clients who accept her abilities as more than just fun. And that isn't funny.

GRAVIOLA POPPYCOCK

It's a cancer-killing dynamo! It attacks cancer safely and effectively without extreme nausea, weight loss, and hair loss! It protects your immune system so you avoid deadly infections! You feel stronger and healthier throughout the course of the treatment! It effectively targets and kills malignant cells in twelve types of cancer, including colon, breast, prostate, lung, and pancreatic cancer! It does not harm healthy cells! It is up to 10,000 times stronger in slowing the growth of cancer cells than Adriamycin, a commonly used chemotherapeutic drug! And it is all natural! What is it? Extract of graviola!

Doesn't that sound fantastic? So how come you haven't heard of this miracle? The "spine-chilling" answer to that question, according to the promoters of graviola supplements, "illustrates just how easily our health — and for many, our very lives — are controlled by money and power." Big Pharma, they say, "is doing everything in its power to keep this natural astonishing cancer cure under wraps in order to protect the enormous profits it reaps from its toxic chemotherapy drugs that do little more than poison patients." Aha! So that's why you haven't heard about it. But how come I have? I guess I'm one of the lucky ones who received a brochure from the Health Sciences Institute featuring a captivating article, "Beyond Chemotherapy: New Cancer Killers, Safe As Mother's Milk." I was also informed that I could view a video about this stunning breakthrough, but I had better do it quickly, because it is unclear how long they will be able to prevent Big Pharma from shutting it down. Furthermore, the Health Sciences Institute has also been able to secure a "limited supply of graviola extract grown and harvested by Indigenous people in Brazil."

Needless to say I was intrigued. How did I miss this? Why wasn't this breakthrough trumpeted in medical journals around the world? Is Big Pharma really that effective at silencing the cutting-edge researchers who discovered the miraculous properties of this natural cancer treatment? Well, it turns out that the Health Sciences Institute isn't any kind of institute, and it isn't very scientific. It's a sales outfit.

So what is graviola? A tropical tree. Also known as soursop, custard apple, or Brazilian paw paw, its fruit is a particular favorite in the West Indies and South America, both for its tangy taste and its supposed medicinal properties. If you can think of a condition, chances are that someone will have reported that it can be treated with graviola fruit, or the fruit's juice, or with teas made from the leaves of the tree. There are anecdotes galore about successful treatment of diarrhea, digestive problems, parasite infections, diabetes, asthma, colds, arthritis, high blood pressure, fever, and anxiety.

Whenever I hear such a wide range of claims on behalf of one particular substance, my alarm bells start to chime. The ailments described have a variety of causes and it is most unlikely they all respond to a single intervention. Our bodies just don't work like that. Asthma and diabetes, for example, are unrelated, and require different forms of treatment. With a natural product such as graviola, there is always the argument that it is composed of hundreds of different compounds, and therefore it is possible that it contains some that may treat asthma and others that are effective for diabetes control. Possible, yes. Likely, no.

A more probable explanation is that the reported benefits are due to a blend of wishful thinking, the resolution of self-limiting conditions, unconfirmed anecdotes, and that good old standby, the placebo effect. Dozens of other plants, fruits, and herbs that grow in the Caribbean and South America have as rich histories as "cure-alls" as graviola. But folklore is not evidence, no matter how compelling some individual testimonials may sound. Especially when it comes to serious diseases such as cancer.

As is often the case with such "wonder products," unethical promoters dredge the scientific literature to find some little grains of truth that they can then ferment into a seductive potion. And there is nothing more seductive, and more marketable, than a "secret cancer potion." In this case, those grains of truth were found in some research carried out at Purdue University back in 1997. A number of compounds isolated from soursop, called annonaceous acetogenins, were tested for their ability to kill cancer cells in the laboratory. The focus was on a particular type of cancer cell that was resistant to the effects of common chemotherapy drugs like Adriamycin. Such resistance is not common, but nevertheless it is of academic interest. One of the compounds in graviola, bullatacin was eventually identified as being effective in killing the resistant cancer cells.

Experiments like this are performed around the world on a regular basis, and thousands of compounds with cancer-cell-killing activities have been discovered. But they rarely progress to anything substantive in terms of human treatment. All that such preliminary findings mean is that further studies may be warranted to investigate whether there is any effect in animals. If that can be documented, then human trials may be indicated. But as far as graviola is concerned, nothing further has been published. Never mind human trials, there aren't even any animal trials that have been published. And we don't know whether graviola, when used as a drug, is free of side effects. There have actually been some reports of Parkinson-like symptoms after taking certain graviola extracts.

None of these concerns have prevented the energetic marketing of various graviola products with headlines such as "Deadly Conspiracy Exposed." The conspiracy, of course, is supposed to be by pharmaceutical companies that want to keep us from finding out about the graviola miracle. I think the only conspiracy here that needs to be exposed is the one that unethical marketers engage in when they try to capitalize on the desperation of cancer victims. Soursop may be a delicious fruit, but claims about its ability to cure cancer leave a sour taste in the mouth.

THE HEALING CODE

Picture a man getting up from a chair and proceeding to point at different parts of his body in a seemingly predetermined sequence, sometimes using his index finger, sometimes his middle finger. Are we watching a *Saturday Night Live* skit? No. We're watching a poor soul attempting to rid his body of disease by following specific instructions embodied in an epic piece of work called *The Healing Code*. Depending on the ailment, a different pattern of finger wagging is indicated. No pills to swallow, no supplements to take, no scalpel to fear.

The Healing Code is the brainchild of naturopath Alex Loyd, whose specialty is the new science of "energy medicine." Actually, there's no new science here, just some old bunk. The basic tenet of energy medicine is that the human body is surrounded by an energy field that is prone to being disturbed. And when the field is disturbed, illness is sure to follow. Luckily, though, according to the sages of energy medicine, these perturbations can be mended through some sort of external energetic intervention. In this instance, finger pointing. They are not bothered at all by the fact that nobody has ever shown the existence of any energy field other than heat radiating from the body. Or that no energy emanates from one's fingers.

The particular offshoot of energy medicine that Loyd practices deals with "cellular memory." It seems that the underlying cause of disease, especially cancer, is the "destructive energy pattern that can be brought about by cellular memories." And what might those be? Cellular memory refers to the supposed ability of the cells in the human body to store a person's habits, interests, tastes, and memories. What the mechanism of such storage may be is not explained. But attempts are made to support the argument with claims that organ transplant patients exhibit the habits and talents of the donor. There is no evidence that this is true. But who needs evidence when pseudoscientific mumbo jumbo can be just as convincing to the gullible?

It's always handy to drag out a reference to Einstein. And energy medicine proponents do. Albert Einstein, they say, proved that all matter is controlled by energy. Actually he did no such thing. The "proof" they talk about is the famous $E=mc^2$ equation, which just states that a tiny bit of mass can be converted into a whole lot of energy under some extreme conditions, such as in an atom bomb. This has absolutely nothing to do with any sort of healing energy. Neither does the fact that a kidney stone can be broken up by energy of the right frequency have anything to do with "energy medicine." Yet this notion is commonly used to demonstrate the healing power of energy. Conveniently, they fail to mention that the required energy to smash a kidney stone is produced by complex electronic equipment and the only role played by a finger is to turn the equipment on.

They also chatter on about MRIs and CAT scans relying on energy of specific frequencies, which of course is true, but again this has absolutely nothing to do with repairing a sick body's nonexistent energy field. And using electrical stimulation in the case of cardiac arrest as an argument to show the "electrical nature" of the body and the potential of "energy medicine" is plain nonsense. But it is the following quote that takes the cake in terms of absurdity. "The Healing Codes discovered a mechanism in the body that allows the 'Super Quantum,' described by Paul Pearsall, PhD in his book *The Heart's Code*, to be stimulated for the remote gathering of information, and to stimulate healing. It transfers the conscious intent of the person as an instruction to the 'Super Quantum' pilot of each cell in the body, which then enacts a healing response in that cell." Meaningless gobbledygook.

Cellular memory advocates claim that traumatic experiences imprint themselves into cells where they "act like tiny radio stations transmitting destructive energy patterns causing disease, chronic pain and shutting down the body's immune system." True healing, they claim, cannot be achieved unless these destructive patterns are removed. According to our inspired naturopath, Loyd, once those destructive patterns are removed, the cancer just "melts away every single time." Quite a claim. Quite absurd. But quite marketable to the gullible and the desperate. It's so simple! To be healed, you just use your fingertips to direct healing energy at the body's four healing centers, according to a specific code. Each disease has its own code. There's even one for ALS, or Lou Gehrig's disease. Never mind that there is no record of anyone, anywhere, ever having been cured of this terrible affliction.

The Healing Code is quite astounding. You can even heal other people! Animals too! And you don't even have to wag your fingers at them. You just have to notify your mind that the fingers are now going to boogie for someone else. Then you just point at your own healing centers in the required pattern. The recipients of the healing can be anywhere, even across the world.

Is there any risk? Well, we're told, there may be a bit of nausea as the body has to clear out the "garbage" before it can get well. Gives me nausea without pointing any fingers anywhere. And, we're told, "when a bunch of toxins are dumped as garbage at once, it is quite normal to temporarily feel worse." There sure is some garbage here that should be dumped. It comes in the form of the idea that you can heal yourself by pointing fingers at mythical energy points.

GETTING DOWN TO EARTH

It seems the Earth is a giant pill. It can cure disease. Luckily you don't have to try to swallow it. But you do have to swallow some pretty bizarre "science" about the supposed curative effect. It's all about the Earth being a source of electrons and disease being some sort of "electron deficiency." The thesis is that the reason we see so much chronic disease these days is that we wear synthetic-soled shoes and walk on carpets that insulate us from the ground. Apparently leather soles are not so bad because sweat from the feet permeates the leather with moisture and body salts so the shoe becomes a semiconductor permitting the body to receive electrons. Living or working in high-rises exacerbates the problem because we are further removed from the Earth's supply of electrons. And things are made even worse by the invisible electromagnetic fields generated by cell phones, computers, appliances, and wiring in the walls, all of which contribute to the body's positive electrical charge and its need to neutralize it with a flow of electrons.

The cure is "Earthing," a process that allows the free flow of electrons between the ground and our body. How do we facilitate this? Simple enough. Just walk barefoot, preferably on damp ground or moist grass, since water is a great conductor. Sea water is especially good because its mineral content enhances conduction. So swimming in sea water, dangling your feet in it, or walking on a sandy beach are great ways to ground yourself. Who says so? Dr. Joe Mercola.

If you haven't heard of Mercola, you have not been surfing the waves of questionable health advice on the web. He is an osteopathic physician whose practice now is limited to offering mostly iffy medical guidance on his popular website and selling a variety of dubious products. Apparently, though, he still occasionally uses a stethoscope because he claims he can hear "electrical chatter" that he attributes to the nervous system moving electrons about. I bet physicians get a charge out of that claim.

Needless to say, Mercola has an answer for those of us who run around in shoes all day, don't work underground, and use cell phones and computers. He suggests sleeping on special mats that allow the free flow of electrons through a grounding wire and exercising on yoga mats of a similar design. These wonder products are of course available from Dr. Mercola's website.

Mercola does not claim to have discovered this simple solution to complex health problems, one that apparently has been missed by the mainstream scientific community. His authority is Clinton Ober, who according to Mercola made the discovery that "could end up as groundbreaking as the germ theory." What sort of background does Mr. Ober have to shake the world in this fashion? It seems that at the age of fifteen he had to forgo school to take care of the family farm. Eventually he forged a career in the cable television business, but in 1993 decided to change his life after almost dying from some illness. For four years he traveled around the U.S. in a forty-foot bus until an epic moment occurred as he stared across a bay in Florida.

"As I was asking myself what I should be doing, I automatically wrote on a piece of paper 'become an opposite charge, status quo is the enemy.'" The Earth, he says, was trying to send him a message that "in the modern electrical world with our bodies insulated from ground contact, we are vulnerable to electrical interference as our cells all transmit and receive the vital information that keeps us alive and healthy." This vulnerability could be countered, Ober determined, "by simple contact with the ground to neutralize charges in the body and thus protect the nervous system and the endogenous fields of the body from extraneous electrical interference."

Before long, Ober found the evidence he needed. He was looking at some studies that claimed to have found adverse effects in humans on exposure to electromagnetic radiation but were deemed to be inconclusive because the effects could not be reproduced in animals. He had hit pay dirt! The reason that sheep and baboons had not experienced the same adverse effects as humans, he reasoned, was because the animals were not wearing shoes and were not sleeping in comfortable beds insulated from the earth. They were naturally grounded! Mercola buys this argument and even contributes his own wacky insight. "Animals that live in the wild are not bothered with inflammation, cardiovascular disease, diabetes, arthritis, or even plaque on their teeth. This is why your dog or cat will crawl under the porch and lie on the bare earth if he isn't feeling well." I figure snakes must therefore be an especially healthy species. Maybe that is why snake oil is so popular.

Selling snake oil always works better if some reasonable-sounding scientific mechanism is provided. In this case it is all about the free radicals that form in the body as metabolic byproducts and are implicated in a variety of disease processes as well as in aging. Free radicals are electron deficient and do their dirty work by stealing electrons from other molecules, such as DNA, damaging them and triggering disease. According to Earthers, contact with the ground allows electrons to enter through the foot and satisfy free radicals' hunger for electrons and thus prevents damage to important biomolecules. Silly stuff. As silly as the original argument that walking with synthetic-soled shoes on carpet insulates us from the ground. Actually, walking on a wool carpet with rubber-soled shoes results in the shoes stealing electrons from the wool and transferring them to the body. Just touch a metal doorknob and watch the electrons jump from the body. That should spark some curiosity about the claims of Earthing being grounded in science.

CURE YOUR ARTHRITIS . . . REALLY?

According to the magazine article, Lucy was forty-eight when the pain of rheumatoid arthritis struck. She didn't take it lying down. The usual regimen of non-steroidal anti-inflammatory drugs, cortisone, and methotrexate provided temporary relief, but the side effects were bothersome. So she decided to go the alternative route. Whole hog. The first practitioner she saw told her to cut that out. The hog that is. Also beef, dairy products, potatoes, tomatoes, alcohol, sugar, wheat, citrus fruits, margarine, eggs, and all packaged and refined foods. Why? Because these foods had shown a positive response in "applied kinesiology" testing. The therapist had measured the muscle strength in Lucy's arm after exposing her to these foods and determined that they were "too harsh for her and were sources of allergic reactions." If only allergy testing were that easy! Since the elimination diet provided no relief, Lucy sought out other "health care providers."

What followed was a truly amazing array of treatments and dietary schemes. There were liver detoxifiers, intestinal detoxifiers, anti-parasite formulas, live bacteria treatments, pancreas extracts, chelated minerals, and organic juice regimens. Then there were the herbal preparations. Everything from devil's claw and celery extract to prickly ash and black currant seed oil. Specifics were determined by a battery of tests which included Essential Metabolics Analysis, Adrenal Stress Analysis, Intestinal pH Analysis, Digestive Stool Analysis, and Darkfield microscopy. She underwent acupuncture, tried homeopathy, fasted for days, ate clay, gulped algae, had her amalgam fillings removed, got vitamin B12 shots, and took huge doses of folic acid and vitamin C. I suspect if there were a Guinness record for most treatments tried for arthritis, Lucy would have set it.

I think it goes without saying that most of these treatments are on a very shaky scientific footing. But that does not prevent desperate people from trying them. History has shown that when scientific medicine leaves a vacuum, a host of alternative practitioners rush in to fill the void. They claim to have the answers that have somehow mysteriously eluded mainstream researchers. Such as the benefits of the Harmony Token. This is a colored disk that's worn around the neck and that "resupplies minerals, vitamins and amino acids with the color that has been stripped away by exposure to electromagnetic radiation." It seems our body doesn't recognize these colorless substances and as a consequence our immune system is weakened. The Harmony Disk utilizes 2,800 colors "to rebuild and repair the body at the cellular level and allows victims of rheumatoid arthritis to resume normal lives." There are testimonials galore. Harmony, we are told, even improves gas mileage and reduces emissions in cars. It makes racehorses run faster. It cures migraines. It also makes me wonder about people's sanity.

Lucy could have also tried "The Incredible Proven Natural Miracle Cure That Medical Science Has Never Revealed." Let me reveal the treatment that is the subject of this amazing book. Urine. Your own. All you have to do is take one to two drops a day. Watch the arthritis disappear. In extreme cases it has to be injected. If you're queasy about consuming urine, how about blood? Autohemotherapy involves taking three-quarters of a cup of blood from a patient's vein and mixing it in a copper bowl with one quarter cup of honey and one quarter cup of lemon juice. Just stir and drink. And there even seems to be some scientific backing here. The *Indian Journal of Orthopaedics* reports that a majority of arthritis patients in a study showed reduction in pain and increase in handgrip strength. Maybe they were just afraid to admit that they hadn't improved lest they be subjected to more of the same treatment.

Then there is snake therapy, which is popular in some areas of China. The gallbladder of a living snake is removed and dropped into wine to make an arthritis cure. Most highly prized are venomous snakes like the king cobra. This treatment has an unusual side effect. Escape of the creatures from shops poses a real problem. I guess seeing your mates being stretched and cut open while still alive provides a certain motivation to get out of there.

This form of snake oil may be a hard sell in North America. But Jogging in a Jug has become a huge business. It was the brainchild of an Alabama farmer who was going to lose his farm and needed an idea. It came in the form of his grandmother's recipe for arthritis. Apple cider vinegar! He himself suffered from arthritis and found he improved when he drank the stuff. His financial situation also improved. At least until the U.S. government decided that according to the claims he was making he was selling an unapproved new drug. Thousands of bottles were destroyed and the company was required to send out letters to consumers apprising them of the situation. Jogging in a Jug is not the only apple cider product for which miraculous claims are made. And they all have impressive-sounding testimonials. What they don't have, however, is scientific evidence. Such products cannot be taken seriously until they are studied seriously.

In some cases attempts for serious study of unusual arthritis treatments have been made. Such as the traditional "raisins in gin" remedy. The idea is to soak the raisins in gin for seven days and eat nine daily. A researcher at the University of North Texas has looked into this and claims that people get significant relief. He uses ninety-proof gin and has discovered that soaking the raisins longer makes no difference but increasing the dose does. Maybe it's due to anti-inflammatory compounds in the juniper berries used to make the gin. Or maybe it's the alcohol. He is now up to thirty-six gin-soaked raisins a day and claims patients feel much happier. I bet they do.

They also say they feel better when they wear copper bracelets. Could there be something to this? A study in Australia using anodized aluminum for placebo control examined copper bracelets. There was a statistically significant difference in relief from arthritis in those wearing the bracelets. Weighing the bracelets revealed that about 13 milligrams of copper was dissolved per month, meaning that if this were all absorbed, the copper level of the body would be increased over the usual amount. Since many enzymes involved in tissue maintenance require copper to carry out their work, a tenuous argument on behalf of copper bracelets on a hand can be made. On the other hand, we have no evidence that oral copper supplements help arthritis.

We've still not reached the bottom of that huge barrel filled with alternative remedies for arthritis. We've not broached aromatherapy, imagery ("just picture the pain flowing out of your body into the nearest creek"), arnica poultices, Chinese thunder god vine (don't ask), or bee sting therapy (it seems that beekeepers who get stung an average of 2,000 times a year have less arthritis — maybe they just don't notice it). We haven't looked at borage seed oil, boron, ground ginger, Boswellia, bromelain, reflexology, DMSO, or the Ayurvedic treatment which involves taking yogaraj guggulu three time a day. We haven't looked at these because scientists haven't looked at them enough to come to any reasonable conclusions. And that I'm afraid is the case for so many of the arthritis treatments out there. They offer hope to people and often not much else. Still, we keep looking.

Maybe the answer will come from gamma-Linolenic acid in evening primrose oil or from drinking orange juice laced with purified chicken cartilage. Maybe the secret lies in a blend of tea and cherry juice. Or maybe there just is no secret. Although Lucy would debate that. After three years of taking virtually everything under the sun, including thirty minutes of sun each day, she claims to be free of arthritis. Go figure.

THE SAGA OF URI GELLER

Click! Uri Geller had hung up. I was very disappointed because I had looked forward to a little mental skirmish with the famed Israeli "paranormalist," as he now bills himself. Much to my surprise Geller had agreed to an interview on my radio show. Why my surprise? Simply because Geller normally cropped up on national TV shows like *Oprah Winfrey* and *The View*. But now he was willing to spend an hour with me. I guess these things happen when there's a book to promote and there's money to be made. There was a book, a rather interesting one, called *Uri Geller's Mind Power Kit*. Unfortunately, we never got around to talking about it.

I have had a long fascination with Uri Geller. The connection goes back to the early 1970s when I was a chemistry student at McGill. I received a letter from an uncle in Israel describing a performance he had seen at a nightclub. An entertainer had wowed the crowd with a number of mind-reading effects and then called a young lady up on the stage. He placed a piece of aluminum foil in her hand and suggested that he would heat it up with mental power. Sure enough, within seconds the volunteer let out a scream and dropped the aluminum like a hot potato. My uncle was impressed, but he did have a few thoughts on the matter. You see, he was a chemist. And chemists of course study matter and the changes it undergoes. He was wondering if there was some chemical reaction by means of which this feat could be accomplished and asked me if I had come across anything like it in my studies. As it happened, I had. Although not exactly through my chemical studies.

As a youngster, I had become interested in magic as a hobby. We're not talking about casting spells here, we're talking about pulling rabbits out of hats, vanishing coins, and yes, heating up aluminum foil. There's an old magic trick that relies on secretly introducing a bit of mercuric chloride into a piece of folded aluminum and placing it in a spectator's hand. As if by magic, the aluminum soon becomes too hot to handle. Of course, it isn't due to magic, it's due to an exothermic chemical reaction. I used to do this until I began to study chemistry. Then I quickly put a stop to it. Mercuric chloride is potentially toxic and it isn't wise to handle it. In any case, I communicated this bit of chemical trivia to my uncle, who agreed that this may have been the way the stage stunt was performed. Still, he said, it had been very well done and he remained impressed with the performer. If you ever get a chance to see him, he said, go! His name is Uri Geller. And that was the first time I ever heard the name that would eventually become world famous.

You can imagine then, that when Uri Geller came to Montreal in the early 1970s, I was there in the ballroom of the old Mount Royal Hotel, ready to witness what I thought would be some sort of magical entertainment. I was in for a surprise. There was no "show," not in the traditional sense, anyway. Geller, a charming and good-looking young man, told us he would demonstrate an ability he had first noticed when he was a child. He would deform metal objects just by concentrating on them. He didn't understand why he had this power, but he had it.

As I recall, the evening consisted mostly of Geller telling us what he had done and what he would do, but there wasn't very much doing. He did display a selection of spoons and keys, some of which came from members of the audience. There was a lot of talk, a lot of metal stroking, and suddenly a cry of surprise from Uri. He seemed more shocked than anyone as he triumphantly held a bent spoon aloft. The audience oohed and aahed. I wasn't all that impressed, because although Geller had kept telling us "it's bending, it's bending" as he was handling the spoon, I really hadn't detected the cutlery's gymnastics. I did see a spoon that was not bent at the beginning of the manipulation and was bent at the end. But there was much commotion and much misdirection, a practice I had become familiar with through my own attempts at prestidigitation. Misdirection is basically the ability to get the audience to look at the wrong place at the right time. By the time it came to twisting keys and starting stopped watches, I was paying pretty close attention and started to get a handle on what was happening. I had come expecting to see some clever deceptions. But not of this variety. I had not expected to be told that the effects we were witnessing were done by purely psychic means without any trickery.

I began to follow Uri's career closely. I watched his TV appearances, I read his books. And I read about him. James Randi's outstanding work *The Truth about Uri Geller* filled in the blanks I had missed and provided explanations for how the now famous psychic could have accomplished all the miracles he performed, had he chosen to do them through tried and tested conjuring tricks instead of through paranormal means. As I became more and more involved with science, Uri's suggestions of special powers began to bother me more and more. Energy can of course be converted from one form to the other, but it cannot be created or destroyed. It takes a lot of energy to bend a key or a spoon. Where was this energy coming from? Uri attempts to address this in *Mind Power Kit* by telling us that coal has hidden energy that is only released when it is lit. Similarly, he says, thoughts and feelings can also generate energy. Yes, there is some truth to that. But here we are not talking about a combustion process. Brain waves can only be measured with sophisticated instruments and the electrical energy associated with these is minute in comparison to the energy required to bend a spoon. We can't even bend a spoon by putting it next to high-voltage power lines!

You can see why I was eager to speak with Uri Geller, especially since he seemed to have reinvented himself as a "healing facilitator." I wanted to give him some credit for getting away from mangling spoons and concentrating instead on the potential of mind power in healing. I also wanted to commend him for some of the charitable work he has done on behalf of hospitals. But I did want to ask him why he has never claimed the James Randi Educational Foundation's million dollars offered to anyone who can perform a psychic feat under controlled conditions. Just bend a spoon and take home a million dollars to give to sick kids. Isn't it irresponsible not to take advantage of this? Wouldn't it be better than giving out Uri bears, which have crystals (energized by Uri himself) hanging around their neck? And there was so much more I wanted to ask. About stories that he had seen a UFO in the Israeli desert, or how women at his performances sometimes experienced spontaneous erogenous satisfaction, or how he had become an expert on electric fields, cosmic radiation, hormones, drugs, and pollution.

But it was not to be. After I introduced myself and recounted the aluminum story, he plainly said he had never done that. OK, maybe my uncle had seen a Uri Geller impersonator. I doubt it though, because I have managed to find an Israeli newspaper article from 1974 describing Geller's use of the aluminum trick. Anyway, when I then tried to talk about his book and began to pose a question about Iscador, a mistletoe extract described in the book as having an anti-cancer effect, the great

Uri Geller, the man who reads minds, the man who terrorizes cutlery with a glance, the man from whom I was expecting an erudite answer, cranked up his mind power and . . . hung up! Just like Randi told me he would if there was any hint of confrontation. It seems the Amazing Randi was the one with the real psychic powers.

I'LL PASS ON AUTOURINE THERAPY

I've often expressed skepticism about the plethora of beverages being promoted these days that claim to energize, calm, heal, or detox our chemically ravaged bodies. "Don't knock it till you've tried it," I'm often told. So, I've gamely downed glasses of noni juice, goji juice, acai juice, vitamin water, oxygenated water, angel tea, and various homemade concoctions including the Master Cleanse, made by mixing lemon juice, cayenne pepper, and maple syrup as recommended by noted nutritional expert Beyoncé. But I've drawn the line at giving "liquid gold" a shot. When I pee into a cup, it is for sending a sample to a lab to be analyzed for creatinine, blood, proteins, ketones, and glucose, all of which can indicate a problem if present in abnormal amounts. But as far as autourine therapy goes, I opt to pass.

Mercifully I was oblivious to this bizarre yet intriguing practice until 1989, when I received a letter from a woman about a therapy that "heals all human beings' illnesses." She had become persuaded about the effectiveness of this "elixir of life" after reading a document, a copy of which she enclosed for my perusal. Would I please help her, she implored, "to save millions of lives with this product that everybody possesses naturally and which God gave us for a medical purpose?" Well, with millions of lives at stake, I figured I better at least have a look at the "data" I was sent. It didn't take long before, what can I say, I got pissed off.

The introduction went like this: "The human race would benefit immeasurably if the medical profession was ended. The proof will be found by observing in big cities and towns, where with increase in the number of doctors, there has always been an enormous increase in the number of patients suffering from various diseases including cancer, heart disease, tuberculosis and diabetes." The document then proceeds to propose a solution to the misery caused by doctors. Sorrows can be drowned in a daily swig of urine! Not only does this remove waste products and toxins, it stimulates the body's defensive mechanism.

So why has this magic potion not been more widely publicized? Could it be that, as urophagists (that's the technical term for urine drinkers) suggest, doctors and "Big Pharma" have conspired to keep the lifesaving information under wraps because "with urine there is no more need for medication or surgery since it kills illnesses in such a short time that doctors are afraid they will lose their jobs." Well, if this were so, doctors should have been weeded out long ago because people in India and China have been imbibing from the Golden Fountain for at least five thousand years! And lest you think the practice is limited to simpletons, Indian Prime Minister Morarji Desai claimed in a 1978 interview on the American news program *60 Minutes* that drinking urine was the perfect medical solution for the millions of Indians who cannot afford medical treatment. He went on to attribute his own good health to indulging regularly. Apparently he wasn't harmed by the practice, living to the ripe old age of ninety-nine.

Perhaps surprisingly, urine is usually quite safe to consume. Bacteria may be present in the urethra, but unless there is an infection, these are generally washed out in the first few seconds of urination, which is why urine samples for analysis should be taken midstream. As far as recycling urine in a situation where no drinking water is available, well, that's not a good idea. Like seawater, urine has a high mineral content and actually can cause further dehydration.

Dehydration is not an issue when urine is consumed for its supposed medical benefits. But can there really be benefits? Attendees at the World Conference on Urine Therapy think so. Each event has attracted an audience of hundreds, who, with appropriate autourine pee breaks, listen to physicians and scientists give evidence about their clinical work with patients as they "aim to help suffering people understand that urine is not a toxic waste but a wide spectrum healing agent, not matched by any other medication."

Although the speakers' interpretation of what constitutes evidence is rather imaginative, they do address a broad spectrum of topics. At the various meetings there have been discussions of dosage, with some proponents suggesting that four-day-old pee is more potent, others claiming that a drop placed under the tongue with an eye dropper is just as effective, conjuring up some confused analogy to homeopathy. One speaker claimed that urine should be "ionized," and described a homemade contraption powered by a solar panel to impart the therapeutic properties. Another reported on the use of camel urine for some abdominal problems, and there was talk of pigs reaching market size faster if reared on their own fermented urine. And there have been anecdotes galore about people being helped in every imaginable condition including AIDS, allergies, asthma, flu, snake bite, and menopausal symptoms, which apparently are best treated with subcutaneous injection of 1 milliliter of urine once a week for four to six weeks. How is all this supposed to work? We're informed that traces of substances that cause illness are secreted in the urine, and when reintroduced into the body trigger the production of antibodies that fight disease. They also trigger skepticism.

I was particularly intrigued by a report about "Plant Urine," never having considered that plants actually voided, although I can affirm that they have been voided upon. Turns out that plant urine is the water that "the plant with its root system filters and lifts from great depths." I was gratified to learn that "no external water sources or artificially processed water is used, ensuring the water contains no unfavorable memories of artificial processing." Sure wouldn't want to consume psychologically disturbed plant urine. Still, I think I would choose it over "The Nectar of Life."

I wonder if eyebrows would be raised if I attended the next World Conference on Urine Therapy to satisfy my thirst — only for knowledge, of course — and then submitted an expense report. Maybe some administrator would say "urine trouble."

MOONBEAMS

Sometimes it's hard to tell the difference between self-delusional simpletons and clever fraud artists. Consider this conundrum. Suppose you take two million dollars of your own money and build a five-story mobile tower supporting eighty-four mirrors that can be focused on a target. Then you plunk this in the Arizona desert, but not for harnessing solar energy or any other reasonable application. No, Richard Chapin did not build this mirror array to reflect sunlight; it was moonbeams he was after. And it was not energy generation that he had in mind, at least not in the traditional sense.

Chapin, a pleasant-sounding fellow with no science background, invested in this giant reflector to take advantage of the "healing rays" of moonlight. As he tells the story, it all began when a close friend was diagnosed with pancreatic cancer, a horrific disease with a terrible prognosis. Chapin could not just stand by and do nothing, and for some strange and unexplained reason, concluded that moonlight therapy may offer a chance at survival.

Now here was a story that the media could latch onto. A bizarre philanthropic healing venture with new-age overtones! As the Interstellar Light Collector took form in the desert, the media began to report on Chapin's theories about moonlight's healing powers. "Moonlight is different from sunlight in that it has its own chemical makeup and spectrum," Chapin declared, "and using these parts of the spectrum may indeed help our bodies and immune system." Utter madness of course; light has no "chemical makeup" and moonlight is nothing other than reflected sunlight. And there is no scientifically plausible way for moonlight to have any healing effect.

Scientific plausibility has never been a barrier to quackery, which derives its power not from evidence but from hope. When word got out that the moonbeam collector was ready to demonstrate its talents, people began to make their way into the Sonoran Desert to experience the supposed therapeutic effects of the reflected light of the silvery moon. Little wonder. Chapin described how Interstellar Light Applications was "making science fiction into science fact" and how he envisioned "moon-glow infusions for cancer, depression and other ailments." Carefully chosen words so as not to make any outright claims, but obviously enticing for desperate people.

There was no charge for a couple of minutes of basking in the moonlight, although a contribution of ten dollars was welcome. And as made clear on the Interstellar Light Applications website, investors were also welcome. If things went well, perhaps a chain of profitable moonbeam healing installations was in the offing. Like any good businessman, Chapin applied for a patent on his Interstellar Light Collector. The application is for a "celestial light collecting device" and does not make any claims about therapeutic effects. Since the mirrors really do collect celestial light, a case for a patent can be made, but filing a patent application has absolutely nothing to do with the device having any health benefit, a nuance that seems to get lost in news accounts that report on the "patented" technology and the mysterious and magical effects of the focused moonbeams.

And those accounts were captivating. There was the hypnotherapist who was cured of his lifelong asthma, the firefighter who after a few lunar doses lost ninety pounds and then completed an Ironman race, and the patient whose acid reflux resolved after a moonlight experience. But it seems the focused moonlight can even help people who may not be able to make it to the Arizona desert to cavort in front of the Light Collector. How? By energizing crystals placed in its path.

Anyone can then purchase jewelry made from these crystals "infused with intensely concentrated moonlight," priced between $70 and $180. But if you go by some of the testimonials, it seems "the power of this ancient and transformational energy" is well worth the price.

Sally from Tucson thought that it was silly to try to feel energy from crystals until they "sang" to her. Then it was "Wow!" And these amazing crystals can even energize money. A lady in a casino pulled out a $20 bill that had been sitting next to an energized crystal in her purse, and guess what? Within three minutes of putting the bill into a slot machine she won a $1,600 jackpot! But that is nothing compared with the truly amazing effect a moonbeam-treated crystal had on a gentleman who had been suffering from hemorrhoids for two months. The crystal cleared up the problem in just half an hour. And the procedure seems risk-free. You place the crystal in your pocket, not anywhere else.

I must admit that it is kind of fun poking fun at such silliness, but there is a dark and serious side to the ridiculous glorification of moonbeams. And rest assured, ridiculous it is. Aside from the well-known placebo effect, sunlight reflected by the moon has no curative properties. Richard Chapin's friend, who unfortunately passed away before the moonbeam collector was completed, would not have been cured of pancreatic cancer. Neither is anyone else going to be cured of any disease by moon-bathing in front of the Interstellar Light Collector. But how many desperate people will be seduced by Chapin's apparently gallant effort to help humanity and squander their money traveling to the Arizona desert to pursue his moonlit folly? As is the case with other woo-isms, some frantic patients may even forgo possibly effective treatments in favor of one that offers nothing but false hope.

Whether Chapin is a brainy con man or an intellectually bewildered humanitarian doesn't really matter. He is promoting a scientifically baseless idea that misleads people. Believing that moonlight can cure disease is sheer lunacy.

NATURAL FALLACIES

Drinking alkaline water can cure disease. Myth. Wrapping tarnished silver in aluminum foil and immersing it in hot alkaline water can remove the tarnish. Fact. Hot water with lemon juice is an effective "detox." Myth. Heavy metal poisoning can be treated with chelating agents such as ethylenediaminetetraacetic acid (EDTA). Fact. Autourine therapy can ward off disease. Myth. Organic agriculture allows the use of certain pesticides. Fact.

Separating myth from fact is the very essence of science and is the focus of many of my public presentations. It is not rare after a talk for someone to ask me what I think is the most prevalent myth I've had to confront over the years. Without doubt it is that natural substances have some sort of property that makes them superior to synthetic materials, with the corollary being that "natural" treatments as practiced by alternative practitioners such as naturopaths are preferable to the methods of "conventional" science.

Natural most definitely does not equate to safe. Natural coniine in hemlock put a quick end to the life of Socrates. In the eighteenth century a local king in Java executed thirteen unfaithful wives by having them tied to posts and injecting the sap of the upas tree through an incision on the breast. That latex contains antiarin, a potent cardiac glycoside. The death cap mushroom is well named, and tetrodotoxin in puffer fish, atropine in belladonna, or batrachotoxin in poison dart frogs can dispatch people pretty quickly. So can natural strychnine, botulin, or arsenic.

Aflatoxins in natural molds are potent carcinogens, and we are familiar with the effects of natural nicotine, morphine, and alcohol. Then of course there are the various pollens released by plants that annoy us with allergies and the myriad bacteria, viruses, and fungi that conspire to do us in with a host of dreadful diseases. And how about the mosquitoes that spread the natural malaria-causing parasite, the ticks that infect with Lyme disease, the snakes that inject a deadly venom, or the wasps that can double the size of your foot with their sting? The fact is that nature is not benign; even something as pleasant as sunshine can be deadly in the wrong dose. Natural radon gas is a carcinogen, and poison ivy can create a great deal of misery. Visiting a urinal without washing hands after handling hot peppers that harbor natural capsaicin will lead to a very memorable experience. Indeed, we spend a great deal of effort trying to outwit the natural onslaught with synthetic antihistamines, sunscreens, and chemotherapeutic agents. But some promoters of natural therapies also spend a great deal of effort trying to outwit us with pseudoscientific mumbo jumbo capitalizing on the "natural is better" myth.

Take for example the cleverly named dietary supplement 112 Degrees, promoted with the slogan "A new angle on sexual health." The geometric reference is to the angle aspired to by men who suffer from erectile dysfunction. 112 Degrees claims to be a proprietary blend of "all-natural ingredients" that enhance male sexual vitality. While the advertising sounds pretty seductive, it is soft on hard facts. The inventor is a "Dr. Laux," who turns out to be a naturopath, not exactly the pedigree one looks for in a drug developer. He is presented as some sort of globetrotting knight in constant search of the best and safest "all natural" treatments. Yup. How likely is it that someone with a smattering of scientific education is going to find an effective product that has eluded the giant pharmaceutical companies staffed by experts who scour the natural world for active ingredients?

The natural health industry commonly promotes the notion that pharmaceutical companies are not interested in natural products because they cannot be patented. This is not so. The use of a specific natural preparation can be patented just like a synthetic drug. Of course what really matters is not whether some substance is patented or not or whether it is natural or synthetic, but whether there is evidence to back the claims. 112 Degrees claims to be supported by numerous scientific studies. Yes, there are some studies, but they don't actually support the claim of enhanced male vitality. The studies show the product is not carcinogenic, that it has some antioxidant potential, and some ability to inhibit an enzyme that interferes with smooth muscle function. All good, but is there even one study to show that 112 Degrees can help men with erectile dysfunction? None that I can find.

The advertising refers to studies about some of the ingredients. Butea superba root, for example. We are told that it was revered by royalty in the ancient kingdom of Siam for its power as an aphrodisiac. That is about as convincing as the story of ancient Assyrian men dusting their genitals with powdered natural magnetic stones and having their ladies follow suit by sprinkling natural iron filings across their own genitals for some literal attraction.

Then there is the claim that *Tribulus terrestris*, another herbal component, combats fatigue and low libido. No mention is made about how much is contained in 112, but we are reassured that Ayurvedic and early Greek healers used *Tribulus terrestris* as a sexual rejuvenator. One study, never duplicated, showed greater mounting behavior in mice, but there are no human studies that have shown any sort of effect on sexual performance or libido. There has been at least one report of breast growth in a man who took Tribulus as a weight training aid, for which it is in any case ineffective. In sheep, Tribulus has been noted to cause Parkinson's-like effects. Of course none of this is noted in the 112 Degrees documentation. So I think a large degree of skepticism, more than 112 degrees, is to be exercised when looking at the overexuberant and naive promotion on behalf of this product by people who are trying to cash in on the unfounded "natural is better" notion.

THE CURIOUS "SCIENCE" OF OSCILLOCOCCINUM

Take the carcass of a duck and place 35 grams of its liver and 15 grams of its heart in a one-liter bottle filled with a solution of pancreatic juice and glucose. When after forty days the liver and heart have disintegrated, dilute the solution to 100 liters. Repeat this dilution another

199 times, shaking in a specific fashion each time. Then take a small pellet of milk sugar and moisten it with the resulting solution. Package the pellets in a box labeled as "Oscillococcinum" and market it to consumers who wish to prevent or treat the flu homeopathically. Not surprisingly, people, especially those leery of pharmaceutical products, were getting more interested in this purported remedy as the fear of the SARS-CoV-2 virus increased. Can protection from viruses really be found in the liver of a bird?

First of all, let's understand what homeopathy is. Unlike what many think, it is not the generalized treatment of disease with natural substances. Homeopathy was the brainchild of Samuel Hahnemann (1755–1843), a German physician who introduced the idea that "like cures like." A substance that causes symptoms of illness in a healthy person, he maintained, will cure a sick person who suffers from the same symptoms. But to effect the cure, Hahnemann argued, the substance has to be repeatedly diluted and thumped against a leather pillow after each dilution. The greater the dilution, the more powerful the remedy! (Hmmm . . . can you die from an overdose if you forget to take the remedy? Just a thought.) An extreme dilution of an extract of cantharide beetles, for example, was to be an effective treatment for urinary tract infections because a concentrated extract caused a burning sensation in the urethra. Hahnemann carried out numerous such "provings" on his family and friends and came up with a "materia medica," or compendium of homeopathic substances, to use in the treatment of disease. But he did not come up with Oscillococcinum. That invention came from Dr. Joseph Roy, a French physician who served in the French army during World War I.

It was during that war that the Spanish flu took the world by storm, eventually killing about thirty million people, 50,000 or so in Canada. Roy naturally took a great interest in the flu and sought to solve its mysterious cause by examining the blood of victims under the microscope. He described seeing tiny microbes that darted or "oscillated" quickly back and forth. He named these "oscillococci" and claimed that they were also to be found in the blood of patients suffering from diseases as diverse as cancer, tuberculosis, and gonorrhea. This "universal germ," as he called it, was responsible for many illnesses! If these oscillococci were causing the symptoms of disease, Roy concluded, then a homeopathic solution of the same should be curative.

Actually, this doesn't even make sense within the tenets of homeopathy, which would require a demonstration that oscillococci can cause symptoms in a healthy person. No such effect has ever been shown, which comes as no surprise given that nobody else has ever seen Roy's oscillococci. To this day we do not know what Roy actually saw through his primitive microscope, but whatever it was, he also observed it in the liver of the Muscovy duck. Appropriately diluted, Roy therefore claimed, this duck liver would work against cancer, syphilis, scabies, and of course the flu. The other claims have fallen by the wayside, but homeopaths still believe that Oscillococcinum can treat the flu. They have been energized by the finding of viruses in birds, including ducks, which they suggest buttresses the "like cures like" argument, an argument that most scientists find is totally baseless.

Hahnemann was not bothered by the fact that after about twelve of his specified dilutions there was not a single molecule of the original substance remaining, because he knew nothing about molecules. Today, homeopaths have to recognize this fact and explain that the dilutions and shaking leave some sort of imprint on the structure of the water used to carry out the dilutions, and that this altered water somehow drives the disease out of the patient.

Mainstream scientists of course find the notions of increased potency with dilution and associated changes in the structure of water all but impossible to accept and take a very skeptical view of homeopathy. But an implausible mechanism for homeopathy does not mean the practice can be dismissed as ineffective. Perhaps homeopathy operates through some modality that science has not yet discovered. Most unlikely to be sure, but not impossible. The true measure of any medical intervention is whether it can be shown to be effective through controlled trials.

Homeopathy has been with us for over two hundred years and has indeed been subjected to a large number of controlled trials. In 2005, researchers at the University of Bristol, led by Dr. Peter Juni, decided once and for all to assemble the results of these trials and determine the efficacy of homeopathy. They searched through the scientific literature and found 110 proper placebo-controlled studies. Their overall conclusion? "We found an effect for homeopathic therapy which is compatible with a placebo effect." A number of other investigations have come to the same conclusion. An analysis of over 200 studies by the Australian National Health and Medical Research Council in 2015 concluded that homeopathy didn't work better than a placebo. What about the effect of Oscillococcinum specifically? That too has been examined. Dr. A.J. Vickers of the Memorial Sloan Kettering Cancer Center in New York looked at the seven studies of Oscillococcinum listed on PubMed, the website which compiles peer-reviewed scientific publications. His conclusion? "Current evidence does not support a preventative effect of Oscillococcinum-like homeopathic medicines in influenza and influenza-like syndromes." He did, however, find that Oscillococcinum increased the chance of a patient considering the treatment to be effective. So, if a viral infection does strike, I don't think I'll put my faith in extremely diluted duck liver juice. But for anyone who does try it, at least they don't have to worry about Oscillococcinum transmitting any avian virus either, seeing that at homeopathic dilutions it does not contain a single molecule from the original bird.

POPEYE'S FOLLY

Mea culpa. I plead guilty to the crime I often accuse others of committing, namely not checking facts properly! Curiously, I would not have discovered my error had I not been doing some proper fact-checking about claims that a nutritional supplement derived from the root of the maca plant can increase libido and alleviate menopausal problems. While looking into this, I came across an ad from Popeye's, a sports nutrition company that sells maca extract. And then as I searched the web for info about this supplier, I came upon an article by criminologist Dr. Mike Sutton: "Spinach, Iron and Popeye: Ironic Lessons from Biochemistry and History on the Importance of Healthy Eating, Healthy Scepticism and Adequate Citation."

I was immediately intrigued given I had once written an article in which I talked about how cartoonist Elzie Segar attributed the strength Popeye needed to rescue his beloved Olive Oyl from the clutches of the dastardly Bluto to the high iron content of spinach. I described that Segar had actually overstated the vegetable's iron content because of a mistake in the scientific literature. Apparently a German researcher in the late nineteenth century had placed a decimal point in the wrong place, ascribing to spinach a tenfold greater iron content than was actually present. I proceeded to relate how Popeye's fondness for spinach led to a huge increase in sales that resulted in statues honoring the sailor man being erected in spinach-growing areas to commemorate his prodigious consumption of the vegetable. And I pointed out that it all was because of an error in the placement of a decimal point! Alas, as it turns out, the error was mine. There was never any decimal point mistake, Popeye was not responsible for a dramatic increase in spinach sales, and most strikingly, Elzie Segar had never claimed that Popeye's strength was due to iron in spinach!

Where did I get my story? I had done my research, just not well enough. My first stop had been an article published in 1977 by Professor Arnold E. Bender, a man who was no slouch when it came to nutritional research. Bender was Head of Nutrition and Dietetics at Queen Elizabeth College in England and had authored over 150 research publications and 14 texts. His book *Health or Hoax?: The Truth About Health Foods and Diets* is a classic. In his article, Bender described the determination of the iron content of spinach by Dr. E. von Wolff in 1870 and how a subsequent analysis in 1937 by Professor Schupan had found that spinach contained no more iron than any other leafy vegetable. In fact, only one-tenth the amount von Wolff reported! "The fame of spinach appears to have been based on a misplaced decimal point," Bender concluded — without any evidence, as it turns out. Given the author's stellar reputation I saw no need to delve further into the matter, especially when a 1981 article in the *British Medical Journal* by hematologist Dr. T.J. Hamblin apparently confirmed the story.

Hamblin's theme was how "frauds, hoaxes, fakes, and widely popularized mistakes run through the history of science and medicine." He explained that linking Popeye's superhuman strength to spinach was due to a decimal point error, and how because of this error Popeye had single-handedly raised the consumption of spinach by 33 percent. He concluded that as far as iron intake went, Popeye would have been better off chewing on the cans the spinach came in. That, he may actually have been right about. Ironically, as Dr. Sutton's meticulous research shows, Hamblin was perpetuating both the myth of the decimal point error and the belief that Popeye guzzled spinach for its iron content.

Sutton did a remarkable job tracing the confused story of Popeye's link with spinach. He did what all good scientists should do: he checked the facts, if possible by going to the original source. While there was no decimal point error, the German chemists who determined the iron content of spinach in the late 1800s did make some mistakes. There may have been contamination of samples with iron from laboratory equipment, and there was also confusion about whether the iron they found referred to that in fresh spinach or the dried variety. A cup of dried spinach would have much more iron given that fresh spinach has a high water content. The fact, though, is that while early analysis may have overstated the iron content of spinach, the correct value was already well known before Popeye was ever conceived.

So is spinach a good source of iron? Yes and no. A cup of cooked spinach contains about 6.5 milligrams of iron, which is a fair amount considering that an average person needs about 8 milligrams a day.

Premenstrual and pregnant women need 18 and 27 milligrams respectively. A cup of raw spinach has less than 1 milligram because of the high water content. But there's another issue. Spinach is high in oxalic acid, which inhibits iron absorption. Basically, spinach is not a great source of iron. And as far as iron providing extra energy goes, that would only be the case if weakness were due to iron deficiency anemia. Popeye, being a sailor, is unlikely to have suffered from such a deficiency given that seafood is an excellent source of heme iron, the most readily absorbed form.

Now for the real crime that Sutton uncovered. Elzie Segar never made any mention of iron in connection with spinach! His first reference to spinach was in a 1932 strip that shows Popeye munching away and declaring that "spinach is full of vitamin A an' tha's what makes hoomans strong and helty." Segar clearly attributed spinach's value to vitamin A, not iron. He erred a little here, since spinach contains no vitamin A, but does indeed contain a good dose of beta carotene that the body converts into vitamin A. (Interestingly, vitamin A helps mobilize iron from its storage sites, so a deficiency of vitamin A limits the body's ability to use stored iron, resulting in an "apparent" iron deficiency.)

In summary, there was never a decimal point error in the determination of the amount of iron in spinach, the vegetable is not a particularly great source of iron, and Popeye never claimed his strength came from iron in spinach. The moral of the story? Even when it comes to apparently trusted authorities, check the facts! And as far as the claims about maca go, I did some careful checking. Here we go.

A good story can sell a product, especially when it comes to dietary supplements. Talk about some legendary use by Indigenous people, throw in terms like "increased stamina," "improved mood," "natural," and "aphrodisiac," and you are off and running to the marketplace. Maca is a plant grown mostly in Peru and its cooked root, with a composition much like wheat or rice, has a long history as a dietary staple. But it is stories about the enhanced virility of Inca warriors who supposedly downed maca root before going into battle that captured the imagination of supplement manufacturers. Although there is no evidence that the Inca fighters actually did this, legends are often based on kernels of truth.

Couple this with anecdotes of Peruvians eating maca root for energy and improved sexual function, and you have a basis for carrying out studies that may potentially lay the groundwork for marketing. After all, plants are fascinating chemical factories, and it is conceivable that maca may have some biologically active compounds. None have been detected so far, but that is not surprising. It takes a monumental effort to isolate, separate, and identify the hundreds of compounds found in plants, and that is only the beginning. Then comes the even greater challenge of testing candidate compounds for biological activity. That's why when it comes to herbal products, the simplest process is to test crude mixtures.

There have been studies of various maca root preparations, and although not compelling, they are suggestive of some potential benefit. In one small study, men taking 1,500 or 3,000 milligrams per day of powdered root claimed increased sexual desire compared with a placebo. There was no measurable change in sex hormones, and curiously the effect was not dose dependent. Another study in young men showed a slight but significant improvement in erectile dysfunction, and one in postmenopausal women resulted in decreased anxiety and depression and some improvement in sexual function compared with placebo. Again, there were no changes noted in any hormone levels.

As usual with such dietary supplements, the consumer is at the mercy of the manufacturer in terms of product quality. There is no systematic checking by regulators that the product actually contains what it is supposed to contain or whether it harbors lead, cadmium, or arsenic, all of which are possible soil contaminants and capable of ending up in the marketed product. Given that maca is widely consumed as a food, it is unlikely that any of the root powders pose a significant health risk, although headaches, stomach problems, sweating, and sleep disruption have been reported in rare cases. It seems that for people looking for a little boost in stamina and sexual function, a daily dose in the range of 1,500 to 3,000 milligrams of Peruvian ginseng, as maca is sometimes called, is an option. It may actually do something, especially if you think it will. And there you have the facts.

BREATHARIANS AND NUTRITARIANS

Breatharians and nutritarians. Never heard of them? One represents the extent of human folly, while the other is a scientifically legitimate attempt to improve health. A breatharian is a person who, under the proper conditions, can live without food. Who said so? Wiley Brooks, who just happened to be the founder of the Breatharian Institute of America. He is no longer doing any breathing, having left us in 2016. "If food is so good for you, how come the body keeps trying to get rid of it?" asked this mental wizard. "Eating is an acquired habit," Brooks was fond of saying, "all of the constituents we need can be taken from the air we breathe."

Of course one must ascend to a certain spiritual level before one can forgo food, and Mr. Brooks was happy to show us the way. But a consultation with this ascended master, who claims to have not eaten for over thirty years, wasn't cheap. The minimum fee was $10,000! Just think, though, of all the money you can save by not having to buy food for the rest of your life. Brooks apparently had lots of experience in advising others to attain "incredible love, peace and joy" through living on air. You see, he has had past lives as Adam, Zeus, Jesus, John the Baptist, Joseph Smith, and curiously, William Mulholland, an engineer who designed Los Angeles's aqueduct system. Quite a puzzle even for believers in reincarnation given that Jesus and John the Baptist were contemporaries.

You would think that Brooks was a unique looney. He's not. Ellen Greve, who has taken on the name of Jasmuheen, heads the CIA. No, not that CIA! This one, the Cosmic Internet Academy, is in Australia and "offers some unusual solutions to world hunger and health issues." Unusual is right. The solution to world hunger is "pranic nourishment." "Prana" is the universal life force that can provide all. It does seem to provide nicely for Greve, who charges $2,000 for her enlightening seminars about eliminating food except for tea and an occasional bit of chocolate or ice cream when she needs a "taste orgasm."

When the Australian version of *60 Minutes* challenged her to put her breath where her mouth is, that is demonstrate that she can live on an intake of "cosmic particles," she failed miserably. A physician ordered the test stopped because this sage, who had authored a book about a 21-day program that allows the body to stop aging and attain immortality by living solely on light, was on the verge of proving her mortality as her kidneys began to shut down. Her explanation? The polluted city air was void of nutrients.

Jasmuheen lived to see another day after being rehydrated, but the same cannot be said for four unfortunate souls whose deaths from dehydration have been linked to following her zany publications. Jasmuheen explains that such tragedies can occur if you haven't found the light that will nourish you.

Prahlad Jani, an Indian guru, claimed to have found the divine light. He supposedly lived without food or water from 1940 until he passed away in 2020. Jani spent his time living in a cave but emerged twice to be tested by physicians, who reported that he did not eat, drink, urinate, or defecate during a two-week observation period. Their account has not been published and has been dismissed by experts who claim that the guru was in fact not observed at all times. But let's not waste any more breath on the nonsense of breatharianism. While we could all eat less, the intake of food and water for humans is not optional. But of course, magical thinking is. So is fraud.

You won't find any breatharians at the Pyramid Bistro in Aspen, Colorado, but you will find some nutritarians. They're a breath of fresh air when compared to breatharians. Nutritarianism may also sound like some strange cult, but it isn't. There is, however, some worship involved, that of "nutrient density." The expression "nutritarian" was coined by Dr. Joel Fuhrman, a family physician who believes that many diseases can be prevented, or even be cured, by eating nutrient-dense foods. Fuhrman recommends a diet based on the Aggregate Nutrient Density Index, or ANDI. The index compares the nutrients a food contains to its calorie content, and assumes that the higher this ratio, the "healthier" a food is.

Just what do we mean by "nutrients"? Generally food components can be divided into macronutrients and micronutrients. Fats, proteins, and carbohydrates provide the building blocks for our body and also serve as our source of energy. Vitamins, minerals, and numerous other molecules that are present in smaller amounts, but have biological activity, constitute the micronutrients. Antioxidants such as beta carotene in carrots, lycopene in tomatoes, or anthocyanins in blueberries are prime examples.

Among green vegetables, kale, watercress, and bok choy top the ANDI list, while strawberries and blackberries lead the fruit pack. Beans rank high, as do sunflower, sesame, and flaxseeds. So do whole grain oats. The nutritarian diet is not totally vegetarian, but about 90 percent of the content comes from nutrient-rich foods such as fruits, vegetables, seeds, nuts, onions, mushrooms, beans, berries, and leafy greens. Compare this with the standard North American diet, in which only 5 percent of calories derive from these foods.

Fuhrman's official blog is entitled "Disease Proof." A little over the top, I would say. There are also excessively optimistic statements about preventing heart disease and cancer, and suggestions that "you don't have to live the rest of your life in pain or on medication." Interestingly, while the dietary regimen is supposed to maximize micronutrient intake, Fuhrman sells dietary supplements. Still, living by the nutritarian credo is certainly preferable to the usual North American diet, and judging by the reviews I've seen about the fare at the Pyramid Bistro, it can be delicious.

I'm keen to try the flaxseed-spelt gnocchi with tomatoes, English peas, snap peas, mustard greens, sunflower seeds, and aged balsamic, or the lemongrass tofu forbidden rice with ginger and steamed bok choy in a spicy carrot emulsion. Forbidden rice isn't actually illegal. It's called forbidden because nobles in ancient China thought it so valuable that they commandeered all that could be grown for themselves and forbade its consumption by commoners. Maybe a research trip to Aspen is in order. I hear even breatharians go there. Seems the unpolluted air is especially nutritious. I would really like to meet one of these science-defying wonders of nature. But I'm not holding my breath.

BITTERNESS ABOUT SUGAR

The play opened in 1960 at the Sullivan Street Playhouse in Manhattan to one of the most scathing reviews ever published in the *New Yorker*. "It is perfectly awful. It is supremely inept. It is magnificently foolish. It is sublimely, heroically, breathtakingly dreadful. It inspires a sacred terror. It is beyond criticism." Obviously critic Donald Malcolm didn't think much of *The Goose*, the brainchild of playwright Jerome Irving Cohen.

That name likely doesn't ring a bell, but perhaps that of J.I. Rodale does. After all, he was the founder of Rodale Press, publishers of the popular *Prevention*, *Men's Health*, *Women's Health*, *Runner's World*, and *Organic Gardening* magazines. Cohen and Rodale are one and the same person. J.I. thought his birth name of Cohen was too Jewish, and when it came to business ventures, "Rodale" would resonate better with the public. The business ventures he had in mind were a push for organic agriculture and for the prevention of disease instead of puzzling over how to cure it.

He had some good ideas, such as cutting down on fat, sugar, and meat, eschewing tobacco, and emphasizing whole grains. His enthusiasm for dietary supplements was unsupported by evidence, but ads for such supplements certainly supported his publications. Rodale himself took some sixty supplements a day and on occasion even demonstrated their supposed efficacy in a somewhat foolhardy fashion. Once to demonstrate the efficacy of tablets he took to strengthen his bones, he deliberately threw himself down a flight of stairs. He wasn't unnerved by the sarcastic remarks this precipitated, maintaining he stood up to them fine because he took plenty of B vitamins that were good for the nerves. No wonder some called him a crank. Rodale, though, had a witty comeback. "Even the critics admit it takes a crank to turn things."

Turn the crank he did. Sometimes in the right direction, sometimes not. He deserves marks for promoting sustainable agriculture and for recommending eating fish and urging people to eat more berries and nuts. But he also maintained that kelp should be included in the diet because "the absence of hay fever cases in the Orient is due to the fact that the Japanese and Chinese eat liberally of this product." He also thought that wild game was a superfood because it was "free of the taint of chemical fertilizers." Fertilizers do not taint food. He might have had a point with pesticides, but not fertilizers.

The guru recommended that white sugar be dispensed with entirely, an idea that sounds sweet to many experts today. But he tainted his advice by saying that sugar should be replaced by maple syrup. This offers no significant advantage. Neither did his exhortations to include more honey in the diet make sense. Yes, it may have small amounts of antioxidants, but sugar is sugar, whether it comes from sugar beets, sugar cane, honey, or maple syrup.

It was Rodale's disdain for sugar that prompted him to pen *The Goose*, an attempt to promote his nutritional thoughts in a rather unconventional fashion. The play introduces John Gabriel, a juvenile delinquent in a Harlem housing project who runs with a gang and even steals money from his sister that he blows on soda pop, donuts, and the horses. His antics prompt a visit by a social worker who explains that juvenile delinquency is caused by an overindulgence in sugar. Her rationale? Consumption of sugar increased elevenfold from 1900 to 1960, with a corresponding increase in crime rate. A classic confusion of association with cause and effect. Then comes a real "scientific" argument. "If sugar can cause cavities in the teeth," she asks, "what will it do to the brain, which is softer?"

Apparently the delinquent youngster is swayed by this powerful reasoning and decides to kick the sugar habit. He becomes a model youth, quits the gang, and returns his sister's money. Somehow this comes to the attention of "food and bottling interests" and one of their portly representatives shows up to offer John a bribe to get back on the sugar wagon. Apparently still morally weakened by residual sugar in his system, he accepts, and soon reverts to hoodlum behavior. Luckily the angelic social worker reappears and once more convinces him to forgo ice cream, pop and donuts, pulling him back from the brink of disaster. And with that the curtain falls. The audience files out with a memorable line still ringing in their ears: "Do you know mosquitoes never bite diabetics?" Rodale's message was that even mosquitoes are smart enough to stay away from sugar.

The harangue wasn't quite over. Brochures promising "The Proofs of Claims made in the Play Called The Goose" were handed out in the lobby. The reader learned that Hitler could never get enough of his favorite whipped-cream cakes and that he was a sugar drunkard. "Was this what made him a restless, shouting, trigger-brained maniac?" The implied answer was "yes." Blaming Hitler's insanity on sugar was insane, but Rodale did bring up the issue of overconsumption of sugar long before it was fashionable to do so.

J.I. thought that following his dietary regimen and supplement intake would allow him to live to a hundred, unless, as he said, he was "run down by a sugar-crazed taxi driver." He didn't make it to a hundred, but it wasn't a taxi driver that did him in. It was a heart attack. And that happened in 1971 during a taping of *The Dick Cavett Show*, just after Rodale had bragged that thanks to his lifestyle he "had never felt better" in his life. The show never aired.

The Goose laid an egg, and not a golden one. It had a very short run, partly thanks to Malcolm's bitter review, in which he promised to express his gratitude to the cast by assuring them that the secret of their identities was safe with him.

A LOOK AT BRACO THE GAZER

What a clever scheme! There's no overt deception. That's because you don't claim to be able to do anything. You don't preach. You don't offer any sort of philosophy. In fact you don't even talk. You don't touch anyone. You don't sell any potions. You don't use any sleight-of-hand tricks. You don't use any sort of equipment. You don't wear strange clothes. However, you do grow your hair to project an image of a certain biblical figure associated with healing. But you don't call yourself a healer, although you do not object if others do. In fact you do nothing but promise to gaze at people for about seven minutes if they plunk down eight dollars. You are Braco the Gazer. And you are a phenomenon!

Picture this. Thousands of people flood into an auditorium, many looking ill, some hobbling with canes, others in wheelchairs, reminiscent of crowds that flock to faith healers, ready to open up their pocketbooks in return for a few miracles. But in this case there are no promises of miracles. Not directly anyway. As the crowd buzzes with anticipation, the proceedings begin with the session's host welcoming everyone to the meeting with "the healer who doesn't call himself a healer." A nice little legalistic out. Everyone's experience will be different, the audience is told, and "skeptics will become believers." "There should be no specific expectations." But of course there are. People have heard that Braco's silent holistic gift can clarify the mind, vanish pain, and wither tumors. It can also repair stalled cars and stop cats from vomiting.

There are a few instructions before the holy man, who does not claim to be one, appears. Cell phones and other electronic devices must be turned off because they may disrupt Braco's "energy," despite the fact that he himself makes no claim to projecting any such thing. Then a warning. The session is only for people over the age of eighteen, because for youngsters the gazing energy is too powerful. Ditto for women who are more than one trimester along in their pregnancy. That's a curious one, because developmental problems are most likely to be initiated in the first trimester. An exception is made on November 23, Braco's birthday, when families can bring children. Perhaps on that day he tones down the energy that he makes no claim to have.

The host's introductory remarks are followed by a video of an unfortunate skeptic who had been diagnosed with "Agent Orange cancer virus" (a ridiculous and befuddling term) and had attended a previous event with the healer who does not claim to be a healer. The skeptic went home, his idea that this was all bunk confirmed. But two days later, a blood test declared him to be cured! (Must be some blood test, capable of detecting a nonexistent virus.) After a few more words about the importance of being skeptical, and instructions to hold up photos or X-rays of sick people to be cured in absentia by the man who claims no healing ability, the time arrives for the "Silent Gaze."

Braco, the Croatian non-healing healer, had been enthralling massive audiences in Europe for some eighteen years before discovering the greenback pastures of America in 2010. In Europe he usually limits himself to just one gazing session per day, but everything is bigger in America. Here visitors can cycle through the lines of "Braco Gazing" all day long, as long as they pay their entry fee each time. And for this all they get to do is gaze at the gazer. Braco struts onto the stage, long hair flowing, face expressionless. You wouldn't be surprised to hear songs from *Jesus Christ Superstar* bursting from the loudspeakers, but all you hear is some new age music. Body almost motionless, he . . . well . . . gazes. That's what gazers do. They gaze. And gaze. Some of the gazees snicker, others revel in rapture, curiously with their eyes closed. Maybe the magical gaze penetrates eyelids.

After about seven minutes, it's over. He glides off the stage, the room empties, ready to be refilled by a new throng, along with some repeaters who feel they need another dose of healing energy from the man who makes no claim to have any. In the lobby there are testimonials galore about toothaches disappearing, back problems vanishing, and bodies being filled with intense heat. But those who came in wheelchairs leave in them. One lady claims to have been overcome by a "big bubble of love." It is not exactly clear what this means, but she seems to have been "satisfied." There are books and DVDs to buy, as well as jewelry that features a thirteen-pronged star. Again, no claims are made other than that the sun is the symbol of life and the sun is the source, which gives us life, light, and energy. Can't argue with that.

Of course not everyone can get to one of Braco's events. That's no problem, though, because thanks to modern technology you can experience the gaze through live streaming. Although you can watch for free, a contribution is welcome. That's a bargain given what people spend on various dietary supplements, magnets, crystals, pendulums, power bracelets, aerobic oxygen solutions, or homeopathic preparations that are marketed with testimonials virtually identical to those heard from people who have been gazed at by Braco.

I thought I'd give the silent gazer a look. He is a good gazer, I'll give him that. But there was no heat, no infusion of vitality, no sensations of inner peace, no awakenings of consciousness, just some thoughts about what he was thinking about as his gaze delivered its dose of placebo. Maybe I should have stuck with it for the full twenty minutes. Maybe it's a dose-dependent thing. I can't complain, though. Unlike detox foot pads, Kangen Water, or zero-energy healing wands that do not live up to their lofty promises, Braco gives you exactly what you paid for. He will gaze at you for the period of time stated. A clever man. A lot more clever than the folks he gazes at.

CELEBRITIES AND CEREBRAL CLAPTRAP

Suzanne Somers is a former actress with a pretty face and probably firm thighs. After all, she did advertise ThighMaster for years. As far as her smooth skin goes, she offers an electrifying explanation. She uses a gizmo that sends tiny jolts of electricity to give her facial muscles a "workout." Right. Unfortunately Suzanne has had to deal with problems that go beyond her thighs and wrinkles. She has had a bout with breast cancer. But neither her good looks nor her struggles with illness qualify her for donning the mantle of a scientific guru. Yet, that is just what she has become. And guruhood means that Suzanne influences many lives as she sounds off on diets, "bioidentical hormones," nutritional supplements, and alternative cancer treatments. Her success in beating breast cancer, she claims, is linked to injections of mistletoe extract. Never mind that she had a lumpectomy and radiation treatment.

In Suzanne's book *Knockout*, physicians like Dr. Nicholas Gonzalez are placed on a pedestal because they are "curing cancer" through unconventional means. Except that the "cures" are not supported by facts. Gonzalez's regimen of numerous dietary supplements and coffee enemas has actually been tested by the National Center for Alternative and Integrative Health, an organization not exactly averse to alternative therapies. Patients fared more poorly than those on conventional chemotherapy. But for Suzanne, this doctor, who has been reprimanded by the New York State Medical Board for "departing from accepted practice," who was forced to submit to psychological examinations and undergo retraining, and who has lost malpractice suits in which he was accused of negligence in cancer treatment, is a hero to whom we should listen. There is no more listening to him, though, because Dr. Gonzalez died in 2015. His death triggered a conspiracy theory that he was murdered as part of a systematic plot to kill "holistic" practitioners.

In 2010's *Sexy Forever* Suzanne expanded her scientific expertise to gene expression. In an interview about this epic she informs us: "We have five cancer protective genetic switches in our bodies that get turned off by diet and lifestyle. One is turned off by toxins and chemicals; one by poor quality food, i.e., non-organic, pesticide-laden food; one by lack of sleep; one by stress; and one by imbalanced hormones. Now, what women really need to understand is that, first and foremost, to turn back on your protective genetic switches you've got to get the hormone switch turned back on. Imbalanced hormones are a big factor in why women are fat, and when women get fat they get very unhappy."

How does such pseudoscientific blather get Suzanne on talk shows? We are talking about a lady who was insulted when her oncologist asked why she was taking steroids. "Who, me? Never!" Suzanne, it seems, had no idea that the bioidentical hormones she was taking were steroids. And let's not even mention the folly of self-treatment of an estrogen-receptor-positive form of breast cancer with estrogenic hormones, which is exactly what Suzanne's bioidentical hormones are.

Why then do people regard her as an authority on beauty, weight loss, and health? Because she is a celebrity! And we have a culture of celebrity worship. Especially if they look good to boot. Talk show sets have become lecture rooms, books written by celebrities are the new texts, and many learn their science from Professors Somers, Tom Cruise, Demi Moore, Julia Sawalha, Gwyneth Paltrow, Michelle Bachmann, and Alex Reid.

You've probably not heard of Alex Reid, but this martial arts fighter and actor is a big name in Britain. He makes the rounds of the talk shows, dates glamorous models, has been in movies and soap operas, and has even had his own TV show. How does he prepare for fights? Let's listen to his philosophy. "It's actually very good for a man to have unprotected sex as long as he doesn't ejaculate because I believe that all the semen has a lot of nutrition. A tablespoon of semen has your equivalent of steak, eggs, lemons, and oranges. I am reabsorbing it into my body and it makes me go raaaaahh!" Makes me go nuts. Give us a break, Alex!

Let's not even address the physiological and nutritional nonsense, but the bit about unprotected sex is dangerous claptrap. Talking about dangerous claptrap, how about British actress Julia Sawalha's views on traveling to places where malaria is endemic. "I don't get inoculations or take anti-malaria tablets, I take the homeopathic alternative called 'nosodes' and I'm the only one who never comes down with anything." Lucky Julia. Perhaps she can point us towards some trials that demonstrate how preparations with no measurable active ingredients can prevent malaria.

Demi Moore also has views on health. She thinks it can be optimized by treatment with leeches. "They have a little enzyme and when they are biting down on you, it gets released in your blood and generally you bleed for quite a bit and then your health is optimized — it detoxifies your blood." Ridiculous. So is Tom Cruise's declaration that "psychiatry doesn't work; when you study the effects it's a crime against humanity." According to Scientology, which Cruise espouses, our problems are caused by mental implants we received from space aliens and should be treated by Scientology's mysterious "auditing" methods. So who is committing the crime against humanity?

Former presidential hopeful and American politician Michelle Bachmann regaled us with her toxicological knowledge. "Carbon dioxide is portrayed as harmful," she says, "but there isn't even one study that shows carbon dioxide is a harmful gas." No, Professor Bachmann, it isn't harmful, we inhale and exhale carbon dioxide all the time. But the issue is the role that carbon dioxide plays in global warming.

Let's conclude with model Heather Mills, who offers insight into our own demise. "Meat sits in your colon for forty years and putrefies and eventually gives you the illness you die of. And that is a fact." No Heather, it isn't.

THE MYTH OF "DETOX"

No wheat, no meat, no dairy, no alcohol, no caffeine, no sugar, no salt, no processed foods. Lots of fruits and vegetables, wheat-free pasta, brown rice, nuts, seeds, beans, lentils, tofu, lemon juice, and liters of water. What do you call such a diet? "Detox," say the food faddists. "Bizarre," say serious nutrition researchers. Detox advocates claim that our modern lifestyle floods the body with toxins, although their definition of this term is somewhat confused. It seems what they mostly have in mind are pesticide residues, food additives (despite the stringent regulations that govern these substances), and environmental pollutants such as PCBs, dioxins, plasticizers, and mercury. But sugar, salt, meat, and dairy also get tossed into the mix as toxic substances. All these toxins, the detoxers claim, build up in our tissues and conspire to bring on weight gain, headaches, bloating, fatigue, lowered immunity, and dull-looking skin, along with any other condition that will get a reader's or listener's attention. We are doomed, they say, unless we periodically flush these toxins from our bodies. And the way to do that is through a detox diet.

Sounds good. But where is the evidence? Has anyone carried out studies to show that "toxins" appear in the urine or feces or sweat after a detox diet? I can't find any such data. The fact is that our bodies are engaged in detoxication all the time. Our liver and kidneys are very adept at removing undesirable intruders, be they synthetic or natural. Is it perhaps possible, though, that a detox diet can increase the efficiency of these organs? After all, there are people who claim they feel better after such a regimen. So, sniffing a potentially hot story, the British Broadcasting Corporation decided to put the detox diet to a test. Producers of the program *The Truth about Food* tracked down ten women aged 19 to 33 who had been partying at a rock festival and were obvious candidates for a detoxification experiment.

Five of the women were put on a classic detox diet, while the others followed a regular healthy diet. All the subjects then sacrificed some of their body fluids for the sake of scientific research. Creatine levels were measured in the urine to monitor kidney function, and blood was tested for liver enzymes to determine the health status of that organ. Blood was also tested for vitamins C and E, indicative of antioxidant potential, as well as for aluminum, which is often targeted as a significant toxin by detox proponents. No significant differences were noted between the groups. There was no apparent detoxification. How then is it that some people claim that they feel rejuvenated after a detox cleanse? Caffeine and alcohol can cause headaches, so eliminating these may be of help. Less food consumed can relieve bloating, and paradoxically, near-starvation can trigger a boost in energy and even feelings of euphoria. This is probably an evolutionary vestige from the times when hungry people had to muster up a last bit of energy in an attempt to locate food.

Even if detox diets do result in improved feelings of well-being, their very concept is flawed. The message is that our body will forgive our dietary sins if we periodically undergo a cleanse. That's not what sound nutrition is all about. The focus should be on eating in a healthy fashion all the time, not on making some dramatic alteration when a problem arises. But that idea doesn't sell nearly as well as claims of health being miraculously restored by a change in diet. The dramatic tale told by anesthesiologist Anthony Sattilaro in his bestseller *Recalled by Life* is a case in point.

Dr. Sattilaro was diagnosed with widespread cancer back in the late 1970s. As luck would have it, he picked up a hitchhiker who had just graduated from a natural cooking school. "You know, Doc, you don't have to die," the man told him, "cancer isn't that hard to cure!" And thus began Sattilaro's plunge into the world of macrobiotics. Desperate people will do desperate things. So out went meat, dairy products, fruit, oil, and eggs; in came brown rice, boiled vegetables, black seaweed, miso soup, and pickled plums.

Almost instantly pain that had been controlled with heavy-duty drugs disappeared, and within three years so apparently did the cancer. *Recalled by Life* became a bestseller and launched numerous cancer patients down the hopeful path of macrobiotics. Needless to say, patients who followed in Sattilaro's footsteps but had no reversal of their fortunes did not end up writing books about their experience. Alas, Sattilaro's cancer returned and this time no diet was able to save him. Did the "detoxing" macrobiotic diet cause the original turnaround? Who knows. Sattilaro also had surgery to remove his testes, prostate, and a rib, and received estrogen therapy.

Certainly Dr. Sattilaro was not the first, nor the last, to claim to have found the secret of restoring health by detoxifying the body. In the 1950s Adolphus Hohensee urged people to insert a clove of garlic into their rectum every evening to rid the body of toxins, and suggested that the scent of garlic on the morning breath was proof that the detoxifying chemicals had worked their way through the body. In the '70s Durk Pearson and Sandy Shaw, "leading independent experts in anti-aging research and brain biochemistry," in their bestseller *Life Extension* urged us to consume some thirty dietary supplements a day. David Steinman came along in the 1980s with his *Diet for a Poisoned Planet* recommending megadoses of niacin to counter pesticides and industrial chemicals in our food.

The '90s brought us Harvey and Marilyn Diamond's *Fit for Life*, which claimed that not eating starches and proteins together was an important detoxification step. The first decade of this century introduced us to naturopath Peter D'Adamo's ideas in *Eating Right for Your Blood Type*. Women with type A blood and a history of breast cancer can benefit from eating snails, he suggests. Alex Jamieson in her *The Great American Detox Diet* (she's the girlfriend who restored Morgan Spurlock to health after he supersized himself by eating exclusively at McDonald's for a month) reminds us of art class, where we made papier-mâché with flour and water. This is just like the goo that forms in our body if we eat white bread, she claims. No white bread for Alex, but lots of sea veggies that can clean out the body. Yummy. I can only hope that the next detox scheme that emerges, and I assure you there will be one, is more tasty, both to the palate and the mind.

As for me, I think I'll forget the gimmicks and abide by the scientific research. I'll have my oatmeal in the morning, sprinkled with ground flaxseeds, topped with berries, and washed down with orange juice. I'll have a grilled chicken sandwich on whole grain bread for lunch with some cherry tomatoes and a banana and a pear. Snacks? Unsalted nuts, carrot sticks, cheese, or live culture yogurt. Beverages? Water, espresso, or tea. For supper, bean and barley soup, spinach salad, followed by my newly developed broccoli, tomato, and rice casserole. Dessert? Strawberries and grapes. And then I'll go to sleep and dream of a smoked meat sandwich, french fries, and a dill pickle. (Occasionally I'll even make this dream come true.) Oh yes, something I forgot, something that I have every day: an apple.

DOUBLE HELIX WATER

Unfortunately chemistry is a mystery to many. And that suits the hucksters just fine. It sets the stage for cashing in on chemical ignorance by bamboozling people with scientific-sounding balderdash. Ignorance, though, is not total. There is one molecular formula that people do tend to recognize, and that is good old H_2O. Then if you press them to name an important chemical in the body, chances are they will come up with DNA. And they are likely to have some sort of mental picture of the double helix structure of DNA, since after all, it's been widely featured in popular books, movies, and TV shows. You can hardly miss the huge model of DNA on the set of *The Big Bang Theory*!

Given that both water and DNA are generally recognized as essential to life, it comes as little surprise that bottles of Double Helix Water have appeared on the scene. The label lists "pure water" as the only ingredient, but does feature a reference to a publication in a physics journal about "stable water clusters," followed by the disclaimer that "the company does not endorse claims or have scientific proof that stable water clusters are effective in the cure, mitigation, treatment, or prevention of disease." Obviously, though, the intent is to imply health claims. Why else would the Double Helix website feature testimonials about improved energy, reduced pain, better sleep, and improved mental clarity? And why would a book by the dynamic duo who promote this wonder water feature the question "Could this discovery save your life?" on the book's cover? And why would that book have a series of pictures that purport to be infrared images of cancer patients before and after drinking double helix water? I have no idea where those pictures actually come from, but the implication is that this preposterous product has some sort of effect on cancer.

Whatever the power of Double Helix water is, it must be potent. Why? Because the water has to be diluted to be used. Imagine the nonsense of diluting water with water. According to the instructions you just add three to four drops to a glass of distilled water and then drink two glasses a day. The mumbo jumbo that explains how Double Helix water works is astounding. Here's a gem: "Stable Water Clusters found in Double Helix Water may act as the body's fundamental building block; a foundation for self-healing and protection from environmental toxins. The water may help bypass blocked Meridians and allow Qi to flow again." And of course it may not. Meridians are mythical channels through which the mythical qi energy flows.

What about the notion that this "newly discovered phase of water can unravel the differences between allopathic and homeopathic medicine." Just what is this newly discovered phase of water? A figment of the promoters' imaginations. Water molecules do form associations with each other, because the partial positive charge on the hydrogen atom is attracted to the negative charge on the oxygen. At any given moment these associations may be described as a "cluster," but they only have a transitory existence, on the order of picoseconds, before the molecules rearrange to another "cluster." These clusters have no observable properties and cannot be "stabilized." Needless to say they have nothing to do with any crackpot ideas of transporting toxins or building bridges over energy blockages. All that you get for forking over about sixty dollars for fifteen milliliters of very ordinary water is a spectacular lesson in hucksterism.

It seems that it would be hard to outdo the double helix water malarkey, but TC Energy Design gives it a valiant try. The real problem with water, you see, is that it is "weakened by flowing in straight pipes and by the unnatural high water pressure." That's why before we drink it, it should be "energized and revitalized." "Consuming food and water of higher vibration supports you both consciously and unconsciously" and "vitalized water supports the purification processes of the body which is essential for health and well-being." And how do you get your water to vibrate properly? Simplicity itself. Just store it for three minutes in a carafe or glass created by Austrian composer Thomas Chochola, the "TC" of Energy Design. This is no ordinary glassware. Oh nooo, it is balanced, harmonic glassware! Chochola has managed to "convert his musical compositions into spatial dimensions using mathematical calculations."

"The shape of the glassware," we are told, "generates an energizing resonance pattern that restores the water within and improves the surrounding environment with subtle waves of harmonic resonance." Needless to say, "all dimensions are musically fine-tuned with one another and with a 6-wave primary structure they emit a major triad, which can be mathematically expressed as a relationship of 1:3:5:8. These ratios can be observed in nature and stand in resonance to superordinate motion sequences in the cosmos. The engagement of the TC shapes with biological naturally occurring factors can be physically described as a coherence phenomenon." It can also be described as incoherent poppycock.

The TC website even has a page pompously titled "The Science." Here we find pictures of "water crystals" before and after the water is stored in the magical carafe. Never mind that there is no such thing as a water crystal. There are of course ice crystals, but no water crystals. The "biological valency" of the water also improves. This, I learned, is a measurement system "used by many dowsers and geomants to locate the vitality of humans and food." Yup, geomants. And what is a geomant? One who practices a method of divination by interpreting markings on the ground or the patterns formed by tossing handfuls of soil, rocks, or sand into the air. If you don't want to take the word of dowsers or geomants, how about the word of a Japanese laboratory that claims to have observed a decrease in stress levels thirty minutes after drinking TC water? I think I need to drink some of the stuff some because my stress level increases just by reading this nonsense.

To bolster the notion that water is affected by its surroundings, proponents of the miraculous carafe refer to Dr. Masaru Emoto's claim that freezing water previously exposed to different words, pictures, or music results in either "visually pleasing" or "ugly" ice crystals, depending on the nature of the exposure. It seems water is even literate, since writing the words "Adolf Hitler" on a glass leads to very ugly crystals. On the other hand, when Emoto's followers focused "feelings of gratitude," whatever that means, on water stored in bottles, the gratitude-focused crystals were rated slightly more "beautiful" than controls.

Dr. Emoto went on to speak to jars of cooked rice immersed in water, saying "Thank you" to one, "You're an idiot" to another, and nothing to the third. After a month, the "Thank you" rice fermented and gave off a pleasant aroma, the "Idiot" rice turned a dark color, and the control produced a disgusting green-blue color. What horror befell the control? According to Emoto, negligence and indifference are the absolute worst things one can do, and he advises us "to converse with children." Not bad advice. Do I really need to mention that Emoto, who passed away in 2014, was not really a doctor? His degree was from the Open International University for Alternative Medicine, a correspondence school in India.

In any case, it is not clear what Emoto's conversations with water have to do with the ability of the TC carafe to produce healing energy or "expand auras." (I suppose nobody wants a constricted aura.) It turns out that to derive such benefits, you don't even have to consume the water. Once you fill the carafe, "the harmonizing effects change the energy of the environment where the carafes are placed." It seems this effect can be enhanced by playing "sacred music" to the water in the carafe. Better yet, if you drink the musical water, the healing frequencies of the music will be carried through the body.

When I expressed skepticism to the promoter of this marvelous carafe, I was flooded with testimonials ranging from cancer patients who had experienced remarkable benefits to a mom who claimed that her three-year-old never asked for chocolate milk again after tasting TC water. I was urged not to knock it before trying it.

Well, I'm knocking. We do not live in a scientific vacuum. We do not concoct ways to trap the tooth fairy. We have enough accumulated knowledge about the workings of the world to know that the shape of a container cannot impart any form of therapeutic effect to water.

Although the carafes designed by Thomas Chochola are pricey, a glass drinking straw created on the basis of "modern quantum physics as well as ancient insights into the natural flow of energy" is available for about $20. Of course it "incorporates the principle of the vortex for an increase in energy level." Just imagine the astounding benefits of sucking double helix water through this straw! Suckers are welcome to give it a shot.

DUNNING–KRUGER

It happened in Pittsburgh in 1995. The bank robber got away with his heist and was totally shocked when he was tracked down by the police the same evening. They had the evidence, the officers explained, brandishing the bank's video surveillance tapes! "But I wore the juice," the befuddled robber was heard to mumble as he was led away in handcuffs. The "juice" he was talking about was lemon juice that he had rubbed on his face, convinced it would make his features invisible to the video cameras. Of course, it didn't do that. What it did do was make his scientific ignorance visible.

Apparently, our robber, obviously not the sharpest knife in the drawer, had read something about lemon juice being used as invisible ink and had totally misunderstood the concept. Indeed, anything written with lemon juice on paper is invisible until heat is applied, turning the writing brown. The chemistry here is quite complex and involves a number of reactions. Cellulose, the main component of paper, is a polymer composed of glucose molecules linked together in a long chain. Citric and ascorbic acids in the juice degrade some of the cellulose to glucose, which then caramelizes with heat.

Caramelization begins with simple carbohydrates such as glucose releasing water, the "hydrate" component of carbohydrates. This leaves behind a mix of simpler compounds that then link together to form brown polymers. At the same time, more dark compounds form via the Maillard reaction as some of the glucose reacts with amino acids in the juice. The same reaction is also responsible for the brown color of roasted coffee, seared steaks, bagels, toast, and french fries.

To complicate things even further, ascorbic acid, or vitamin C, also undergoes changes with heat to yield furfural, a compound that can go on to form colored polymers. Furthermore, furfural can also engage with amino acids in the Maillard reaction. Clearly, the reaction of lemon juice with heat presents some very complicated and interesting chemistry, but it suffices to say that heat will reveal the presence of any "invisible" marks on paper made with lemon juice. Obviously lemon juice does not render anything "invisible."

The scriptwriters of *National Treasure*, an entertaining movie, also showed their chemical incompetence in a scene in which the heroes visualize some secret writing on the Declaration of Independence by spreading lemon juice on the document and blowing on it with supposedly hot breath. This is a total misinterpretation of the role of lemon juice in invisible writing. The only possibility, in this case, is the document turning brown wherever the juice had been applied, obliterating any secret message. Never mind that breath is not hot enough to change the color of lemon juice.

The somewhat dimwitted bank robber's belief that he had come up with an ingenious scheme caught the attention of psychologist David Dunning at Cornell University. Dunning had been interested in the subject of how people's assessments of their intellectual abilities compare with objective evidence of their mental prowess. Indeed, the doomed bank robber had believed himself to be more clever than he actually was. To further explore this phenomenon, Dunning and his graduate student Justin Kruger designed a test based on simple general knowledge that they administered to undergraduate students, who were then asked to estimate how well they did and how they thought they would rank when compared with other students. Amazingly, the students who did the most poorly on the test were the ones who most significantly overestimated their success and believed they had performed better than others.

To see if such overestimation of knowledge extended beyond the classroom, the Cornell scientists questioned gun hobbyists about safety. The ones who got the fewest questions right were the ones who overestimated their correct responses the most. Such lack of awareness of incompetence has since been referred to as the Dunning–Kruger effect, first formulated in 1999. Many others throughout history have, however, recognized the problem of some people's inability to recognize the true level of their knowledge, leading to an overestimation of their competence.

Confucius remarked that "real knowledge is to know the extent of one's ignorance," and in *As You Like It*, Shakespeare noted, "The fool doth think he is wise, but the wise man knows himself to be a fool." Alexander Pope told us that "a little learning is a dangerous thing" and Charles Darwin opined that "ignorance more frequently begets confidence than does knowledge." Famed philosopher Bertrand Russell once noted that "one of the painful things about our time is that those who feel certainty are stupid, and those with any imagination and understanding are filled with doubt and indecision." Russell's remarks ring true today more than ever, with various bloggers with no relevant background pontificating in a simple-minded manner on complex issues such as cancer, genetic modification, and diets. The Dunning–Kruger effect is indeed in full bloom.

MEDICAL MEDIUM

Celery growers are thrilled. The vegetable is selling like hot cakes. There have been runs on celery before, usually coinciding with the appearance of some article claiming that more calories are needed to digest this vegetable than it provides. While it is true that celery contains only about six calories per stalk, there is no food that results in "negative calories." But that bit of nonsense is nothing compared with the truckload of detritus that is being dumped on consumers by promoters of celery juice who claim that it is a veritable cure-all. The "brains" behind this puffery, and the man responsible for the boost in celery sales, is Anthony William, who calls himself the "Medical Medium."

William freely admits that he has no medical education of any kind, but then again he doesn't need any. That's because he gets his information from an all-knowing spirit, who somewhat unimaginatively is actually called "Spirit" and speaks to William in a clear voice that only he can hear. Let's have William tell the story himself, as he does in his book *Medical Medium*, which, believe it or not, made the *New York Times* bestseller list.

"My story begins when I'm four-years-old. As I'm waking up one Sunday morning, I hear an elderly man speaking. His voice is just outside my right ear. It is very clear. He says 'I am the Spirit of the Most High. There is no spirit above me but God.' Later that evening I suddenly see a strange man standing behind my grandmother. He has gray hair and a gray beard, and is wearing a brown robe. When none of my family reacts to his presence, I slowly realize that I'm the only one who sees him. He says 'I am here for you.' Then the gray man looks at me, 'now say, Grandma has lung cancer.'"

William then recounts how his grandmother was shaken by this bit of information coming from a four-year-old, and even though she felt fine, she made an appointment for a general checkup. A chest X-ray revealed that she had lung cancer! And so with the help of a man that only he sees, speaking with a voice that only he hears, at the young age of four, a career as a medium with a talent for diagnosing and treating disease was forged.

When you open *Medical Medium*, subtitled *Secrets Behind Chronic and Mystery Illness and How to Finally Heal*, to chapter one, you can feast your eyes on the first sentence. "In this book, I reveal truths you won't learn anywhere else. You won't hear them from your doctor, read them in other books, or find them on the web." I think that is true. According to William, the reason we don't hear of the "truths" elsewhere is because Spirit has "insights into health that are decades ahead of what is known by medical communities."

Spirit also teaches William to perform diagnostic body scans. His training though is not in hospitals but in cemeteries. "I spent years in different cemeteries performing this exercise with hundreds of corpses. I became so good at it that I can almost instantly sense if someone's died of a heart attack, stroke, cancer, liver disease, car accident, suicide or murder." Quite a gift. Although it seems the technique may need a bit of work, judging by William giving an "all clear" diagnosis to a TV host who was soon after diagnosed with malignant myeloma. William doesn't need to be in the same room with a person to do a reading, he can do it by phone as he demonstrates on his radio show. He asks callers for their symptoms, and then diagnoses them, usually as suffering from an infection by the Epstein-Barr virus, and prescribes a treatment, which often is sixteen ounces of celery juice. Some would call this practicing medicine without a license.

There is no such thing as an autoimmune disease, William says, and multiple sclerosis is a version of the Epstein-Barr virus that can be cured with a host of supplements such as barley juice extract powder. Attention Deficit Hyperactivity Disorder (ADHD) as well as Lyme disease (which William absurdly claims is a viral infection) can be treated with supplements, and digestive problems respond to, guess what, celery juice. Indigestion, we are told, is a problem because the six components of hydrochloric acid in the stomach (a ludicrous notion) don't work well together. Celery juice works its magic because it contains "unique sodium compositions." Spirit seems to be in need of a chemistry lesson.

William recognizes that there are problems with plastics, but Spirit has a solution for this as well. "Anti-Plastic Tea: is a blend of equal parts of fenugreek, mullein leaf, olive leaf and lemon balm." This is different from Anti-Radiation Tea, a blend of Atlantic kelp, Atlantic dulse, dandelion leaf and nettle leaf. William can also diagnose, with Spirit's help of course, Alzheimer's disease. When a lady was exhibiting memory problems, William scanned her and found "two large pockets of mercury in the left hemisphere of her brain." A heavy metal detox regimen with barley grass juice extract powder, spirulina (preferably from Hawaii), cilantro, wild blueberries (only from Maine), and Atlantic dulse was the prescription.

Incidentally, Spirit is also a talented car mechanic, advising William on fixing "unfixable" cars to the utter shock of legitimate car mechanics. And oh yes, William also sees angels. One helped him save his drowning dog. There are twenty-one essential angels, he tells us, and they can be called upon for help, but the call must be out loud and the specific angel has to be named. You can call upon the Angel of Purpose if you are struggling with your purpose on Earth and you can ask the Angel of Water to change the frequency of the water you bathe in to make it more cleansing, nourishing and grounding. There are also exactly 144,000 "Unknown Angels" who don't have names but "are eager for the chance to work with us and by summoning them we can tap into a resource of profound power for healing the body, mind, heart, spirit and soul."

And if the angels aren't up to the task, you can go for the birds and the bees. "Birds sing the song of the angels and the heavens and can mend fractured souls and reverse disease." How does that happen? "The frequency of these melodies resonates deep within our DNA, which allows it to reconstruct the body on a cell level." If you don't care to listen to birds, you can watch bees. "As bees dance from flower to flower, they emit a healing frequency that reverses disease and promotes soul and emotional restoration."

If you are not keen about listening to birds or watching the bees and are not comfortable calling out loud to angels, there's always celery juice. There is nothing special about this juice. Like any vegetable, celery does contain a host of chemicals, notably apigenin and luteolin, that have antioxidant, anti-inflammatory and anticancer properties. However, these effects have only been observed in rodents and cell cultures using the isolated compounds, not the vegetable or its juice. There are hundreds of compounds found in fruits and vegetables that have such properties, which is why the best bet is to eat a wide selection. There is no single "superfood" or beverage.

Celery juice does have one practical use. It is a source of nitrates and nitrites and is now used in processed meats so they can declare "no added nitrate or nitrite" on the label. Nitrates and nitrites are a concern because of potential conversion in the body into carcinogenic nitrosamines. Of course, the source of nitrite is irrelevant. Whether it comes from celery juice or from a bottle purchased from a chemical company makes no difference. For a 150-pound person, 16 ounces of celery juice provides roughly three times the daily recommended dose of these chemicals. So drinking all that celery juice is not such a great idea. Apparently Spirit can also use a course in food science. Maybe in economics as well.

Depending on the cost of celery, which does vary seasonally, it can cost from four to six dollars a day to produce the amount of juice that William recommends. If you don't want to squeeze it yourself, it is available commercially for around $17 for 200 milliliters. That works out to about $40 per day if you follow William's protocol. Or you can buy celery juice powder for $165 for 500 grams. The packaging says "contents may do you good." Or they may not.

What can we conclude about the "Medical Medium"? There are only three possibilities. One, spirits exist, and there is at least one with medical knowledge that is superior to that possessed by any physician and has selected Anthony William to impart this knowledge to ailing people. Two, William is a clever huckster who capitalizes on people's desperation. Three, William is a candidate for psychiatric care. You decide.

FITTING SQUARE PEGS INTO ROUND HOLES

Stephanie Seneff is an MIT computer scientist with no expertise in agriculture, chemistry, toxicology, or the biological sciences. For some reason she has decided that genetically modified foods are the tools of the devil and that glyphosate (Roundup), the herbicide used to kill weeds in fields of crops that have been genetically modified to resist it, is responsible for many of society's ailments. She claims that exposure to glyphosate causes autism and attempts to prove this by flashing a graph showing that an increase in the incidence of autism parallels an increase in the sales of glyphosate. Confusing correlation with cause and effect is such a fundamental error that anyone who builds a case in this fashion can immediately be discounted as a reliable scientist. Autism rates have also increased in step with an increase in vegetarian diets, yet nobody is claiming that vegetables cause autism. Boating accidents correlate with the sales of ice cream, but ice cream does not cause the accidents. Clearly, more ice cream is sold in the summer and there are more boating accidents in the summer.

For years Seneff has tried to make a global name for herself by becoming the people's champion, a knight clad in shining armor, riding a white horse into battle against an agricultural industry that only cares about profits. She has been widely criticized for her abuse of science but a recent diatribe is truly comedic. She is riding the coattails of COVID-19 in a ludicrous attempt to implicate glyphosate as a factor in the disease. How? By trying to forge an absurd connection between biofuels and COVID-19! The biofuel in question is ethanol, produced by the fermentation of plant products, mostly corn, and is added to gasoline and aviation fuel to reduce reliance on petroleum.

Ethanol is not only a renewable resource, it also burns more cleanly than gasoline, reducing air pollution.

Seneff's thesis is that the genetically modified corn used to produce ethanol has been sprayed with glyphosate and that the chemical is therefore present in the gasoline to which the ethanol is added. When the fuel burns, the glyphosate is aerosolized, and when inhaled, affects the immune system resulting in the "cytokine storm" that characterizes some cases of COVID-19. What is her evidence? She preposterously introduces the case of a couple in their seventies who both died from the disease. In her words: "The Mars were both in their seventies, so they match the profile of increased susceptibility due to older age. But perhaps a more significant factor was the fact that their restaurant was located just a few blocks from Interstate 5, an eight-lane highway where trucks, buses, and cars passed by all day long, spewing out toxic exhaust fumes."

Glyphosate decomposes at 187°C, a fact that Seneff could easily have looked up. Not only is it nonsensical to suggest that glyphosate survives combustion, there is no evidence that it is ever found in ethanol. Is there any evidence that glyphosate has any effect on adaptive immunity as she claims? Here she brings up the example of a farmer who tried to commit suicide by drinking a cup of a glyphosate-containing herbicide and developed a precipitous drop in blood pressure, along with hypoxia, respiratory distress, and acute pulmonary edema. No great surprise there, but drinking a toxic amount of glyphosate is hardly a model for inhaling trace amounts from the air, should these actually be present.

Then Seneff invites further scientific ridicule by claiming that glyphosate runoff from agricultural fields plays a role in the outcome of coronavirus infection. The Yangtze River runs through Wuhan and is highly polluted from runoff, and this, she claims, results in glyphosate in the surrounding air, which in turn affects the immune system. This is what caused the coronavirus outbreak in Wuhan to have such disastrous effects. Has anyone ever found glyphosate, a non-volatile compound, in the air around water systems? No. She goes on to point the finger at hotspots like New York, New Orleans, and Seattle, arguing that the Hudson River, the Mississippi, and water around Seattle are sources of glyphosate. Never mind that cities with major population densities tend to be located around major waterways and crowding is the reason for transmission of infection.

What about Italy, a country that does not grow genetically modified crops, being a hotspot? Seneff has another asinine explanation here. "Italy has developed a technology that involves gathering used olive oil from restaurants and converting it into biodiesel fuel. While Italy does not allow GMO crops, glyphosate is used routinely to control the weeds growing around the olive trees." So, according to her, the Italians' love of olives plays a role in the spread of COVID-19. And Russia? Fewer cases. Why? Because, according to this champion of science, they don't use biofuels. Never mind that Russian reporting is notoriously suspect.

There is still more. Seneff goes on to stuff another load of bunkum into her giant pile of pseudoscientific detritus. She proposes that the lung problems associated with vaping are also caused by glyphosate. How so? Propylene glycol and glycerol are components of the vaping mixture in e-cigarettes and they are sourced from waste from biofuel production. Indeed, glycerol is a byproduct of biofuel production, but it is also commonly sourced from animal tallow. It can be converted into propylene glycol, although the main production method of propylene glycol is from propene. In any case, no glyphosate has ever been detected in vaping mixtures, but even if it were present, it would of course not survive combustion! As if she had not blown enough hot air, she then marshals vitamin E acetate into the dock. Seneff knows why this compound is suspected of being responsible for some of the lung problems encountered by vaping. What other scientists have missed is that vitamin E is commonly sourced from soybean oil, "probably from GMO 'Roundup Ready' soybeans," and is therefore contaminated with glyphosate! Any evidence for this? Zero.

What we have here is a series of laughable arguments that represent an illness sometimes seen among scientists who are wedded to a theory they have formulated and will try to fit square pegs into round holes to prove they were correct. In this case, in a sense, *that* illness is caused by glyphosate.

BEE POLLEN AND THE OFFICE OF ALTERNATIVE MEDICINE

Why bother going to a doctor if you feel sick? Just walk into any bookstore these days and check out the health section. There is a cure for everything. If you have asthma, just rub oil of oregano on your chest. Digestive problems? You need the Clay Cure. Magnets will relieve your arthritis pain and aromatherapy is the solution for ailments ranging from cystitis to anxiety. Then there are the nutritional regimens. Depending on which book you leaf through, salvation lies in flaxseed, fish oils, garlic, oat bran, soy protein, red wine, freshly squeezed juices, apple cider vinegar, or barley green powder. And let's not forget those supplements. Vitamins, pycnogenol, blue-green algae, tea tree oil, bifido bacteria, natural enzymes, and shark cartilage are all ready to come to the rescue.

If none of this appeals, then you can try drinking some tea brewed with the revolting, slimy kombucha "mushroom" or experiment with colorpuncture, a technique that focuses colored light on acupuncture points and "energizes powerful healing impulses." Uri Geller's Mindpower Kit will tell you how to use crystal quartz for psychic healing and books on feng shui will teach you how to harness positive energy from the environment through the correct placement of furniture and decorative items. You can also discover the secrets of holistic bathing (whatever that is), chelation therapy, bee pollen, homeopathy, Ayurvedic medicine, natural hygiene, chiropractic, catalyst-altered water, colonic lavage, therapeutic touch, coffee enemas, and naturopathy. Confused?

that there is a good chance they can help. They offer hope, although often it turns out to be false hope.

What we really need is a thorough scientific examination of the alternative therapies that show promise based on personal testimonials. This is starting to happen. In the United States, the Office of Alternative Medicine was created in 1991 and given a budget of two million dollars. It was later converted to the National Center for Complementary and Integrative Health (NCCIH) and had a budget of $151.9 million for 2020. The center will even award grants and organize clinical trials. Perhaps we can look forward to some interesting results, but since 1991 not much has happened. Not a single alternative treatment was found to be highly effective and not a single one was completely debunked.

It is surprising that the Office of Alternative Medicine did not thoroughly investigate the potential of bee pollen therapy, since it played a crucial role in its creation. While herbalists promote bee pollen for all sorts of conditions, ranging from asthma to enlarged prostate, and various bee pollen preparations can be found in health food stores, there is no proof of efficacy. The force behind the establishment of the office was Senator Tom Harkin of Iowa. But the spark was provided by another politician from Iowa, Berkley Bedell. This gentleman became thoroughly taken by alternative medicine when he was apparently cured of Lyme disease and prostate cancer by using colostrum, the first milk of a cow that has just given birth. This treatment was the brainstorm of Herb Saunders, a Minnesota dairy farmer, who for $2,500 would sell people a pregnant cow, inject some of the patient's blood into its udder, and then provide them with the colostrum. He claimed that the colostrum "has the power to wham out cancer." The authorities didn't agree. Saunders was twice arrested for swindling, the mistreatment of animals, and practicing medicine without a license. But the colostrum did seem to cure Bedell, who then contacted Harkin, searching for ways to fund alternative medicine.

During their discussions, the subject of Harkin's allergies came up. Bedell, already fancying himself as an expert in alternative care, suggested Harkin try bee pollen. He started taking the tablets, sometimes up to sixty a day, and claimed that after six days his allergies disappeared. Understandably he was very impressed and immediately began to lobby for the establishment of the Office of Alternative Medicine. Today he says he still sometimes has allergies, but just takes more pollen and they disappear. Unfortunately there have been no corroborating studies. Anyone contemplating using bee pollen should also realize that in rare cases it has produced life-threatening allergic reactions.

Even though the cures that stimulated the creation of the Office of Alternative Medicine are suspect, its spin-off, the NCCIH, serves a useful purpose. Scientific investigation of the claims of alternative medicine is sorely needed. It may turn out that colostrum actually has beneficial properties. Some studies have shown that cows may in fact form useful antibodies to injected microorganisms. Even bee pollen may turn out to be useful. But what we need are facts, not hype. In the meantime, I still despair when I walk through the health aisle in the bookstore, because I wonder how many people are unsuccessfully trying to restore their health with coffee enemas or autourine therapy. Maybe I'm just being crabby. Maybe what I need is an extract of Malus pumilia (crab apple), which according to the Bach flower therapy book reduces despondency and increases broad-mindedness.

SPOONK

Some time ago I spent hours hammering hundreds of long nails through a plank of plywood. Wasn't easy. The nails had to be carefully spaced, about a centimeter apart, protruding exactly the same distance from the wood. Any deviation would have made it quite uncomfortable to lie on my new bed of nails! The point, as it were, was to demonstrate to students that you did not have to be a Hindu mystic to lie on a bed of nails. There was absolutely nothing paranormal about accomplishing the feat. It was simply a question of physics. As long as the weight was distributed over enough nails, there was no worry about skin penetration.

While this was a neat demonstration, I can't say it was particularly relaxing. That's why I was taken a little aback when I came across what amounted to a bed of nails being sold in a health food store with claims of promoting relaxation and stress reduction. Not only that, it promised to energize, improve sleep, reduce pain, increase circulation, and within five minutes provide a fresh glow and face-lift effect. Meet Spoonk, the "acupressure massage eco mat"!

Spoonk is a flexible plastic mat, which instead of nails features 6,210 sharp plastic stimulation points. The odd name is a whimsical version of "spunk," a word made up by Pippi Longstocking, the fictional heroine created by Swedish children's writer Astrid Lindgren. While Pippi attached no meaning to the word, it became associated with strength, energy, and a love of life, all characteristics Pippi possessed. The apparent message is that Spoonk can bestow these very properties. The rationale is based on the concept that the body is permeated with channels called meridians through which a sort of life energy, often referred to as qi, flows. Any blockage of the flow of qi means bad news.

According to traditional Chinese medicine, these blockages can be cleared either by the appropriate application of needles, as in acupuncture, or with physical pressure. Such "acupressure" is said to be the principle behind Spoonk. The little spikes are designed to produce some sort of a shotgun effect, clearing all the possible meridian blockages. Anatomical science, however, cannot detect any sort of meridian and no measurable qi force exists. Of course that does not mean that acupressure cannot work by some other means.

My first encounter with "acupressure" was back in the 1960s in an introductory psychology course at McGill, although the term was never mentioned and there was no talk of any qi. I was lucky enough to attend a lecture by Professor Ronald Melzack, one of the world's premier experts on pain and developer of the McGill Pain Questionnaire, used by cancer clinics around the globe. Frankly, I don't think I really understood his "gate theory," and still don't, but I know it has something to do with the spinal cord either blocking pain signals or allowing them to pass to the brain, depending on whether the signal travels via small nerves or through larger nerve fibers. Somehow by applying appropriate pressure in certain spots, the "gate" that allows a pain message to pass to the brain can be blocked. It all sounds very theoretical, but Dr. Melzack's practical example made an impact. If you have a toothache, he advised, just rub your hand between the thumb and forefinger with an ice cube. That sounded pretty odd, but Professor Melzack wasn't talking out of his hat; he actually quoted a clinical trial that had demonstrated success.

Of course rubbing a hand with an ice cube for relief of a toothache is a far cry from relaxing the body by lying on a plastic bed of spikes. Or energizing it. Actually, it isn't clear to me how you can be both energized and relaxed at the same time, but never mind that. The pertinent question is whether there is any evidence that the mat provides benefit. I can't find any trials that have put Spoonk to a test, but I did turn up some sketchy data about a handmade mat with some 1,500 stainless steel office pins, invented by a Russian layman, Ivan Kuznetsov, back in the 1980s. He figured that by poking the body everywhere, he would be hitting some of the right acupuncture points and no harm would be done by any that were off target. The mat was eventually sold in pharmacies and spurred a television documentary that detailed successes in alleviating pain. No studies, however, were published in the scientific literature.

An American version of Kuznetsov's mat was marketed for a while under the name Panacea. In one study, albeit not methodologically impressive, of 200 users, 98 percent reported pain relief, 96 percent reported relaxation, 94 percent reported improvement in the quality of sleep, and 81 percent reported an increase in energy level. Approximately half of the subjects with allergy problems reported relief of their symptoms. The fly in the ointment here is that subjects commonly report such benefits no matter what kind of treatment they are offered. Similar benefits are claimed by people who rub their body with snail slime, drink oxygenated water, sport plastic bracelets with "therapeutic" holograms, or bask in the reflected light of the Arizona moon.

While there are no compelling studies about acupressure mats, there have been a surprising number of studies on various other forms of acupressure. In fact, some forty-three studies have investigated stimulating various body parts to manage pain, breathing problems, fatigue, insomnia, and pregnancy- or chemotherapy-induced nausea and vomiting. While some studies have shown benefit, a systematic review published in the *Journal of Pain and Symptom Management* concluded that "the existing clinical trials do not provide rigorous support for the efficacy of acupressure for symptom management."

Spoonk asks us to "imagine a world free from stress, anxiety, depression, stiffness and pain." The company's stated mission is "to contribute to that imaginary world with a simple and effective product." I can go with the "imaginary," but the use of the term "effective" would have left me somewhat skeptical had I not noted the logo of *The Dr. Oz Show* on Spoonk's package. Surely Dr. Oz would not lead us astray! What a stressful thought. So I dug out my old bed of nails. After all, according to Spoonk, stimulating all those acupressure points is a great stress buster. Didn't work. Tore my pants.

SYLVIA BROWNE

She was rich. She was famous. And she was heartless. Who but an unfeeling person would tell the parents of a missing child, without any evidence, that their son is dead? Sylvia Browne, that's who. "The best psychic reader in the world," as she used to bill herself. Well, maybe not quite the best.

Shawn Hornbeck, the boy Sylvia declared dead on national TV, turned up very much alive, the victim of a kidnapping. The kidnapper was not Hispanic, and did not have dreadlocks as the psychic had suggested. And then there was something else the psychic did not foresee. The devastating effect her words had on the parents. Or maybe she didn't care. After all, such dramatic remarks are great for ratings.

Browne often held court on *The Larry King Show* and the *Montel Williams Show*, where she predicted futures, talked to the dead, and dispensed advice on matters ranging from love to health. This talented seer could just look at someone and tell that they had a difficulty with their prostate or that they should check their bilirubin, a "liver enzyme," as she called it. Of course bilirubin is not a liver enzyme, it is a breakdown product of hemoglobin, the molecule in red blood cells that transports oxygen. But I suppose someone with a degree in English literature, which is what Sylvia had, should not be expected to know that. Maybe, though, Sylvia was not to blame. Much of her prattle, she claimed, actually originated from "Francine," her "spirit guide." And maybe spirits weren't really up on their biochemistry.

I must admit that I get really irritated by self-proclaimed "authentic psychics" who prey upon the gullible. Maybe that's because I don't believe the dead talk to us, that our futures can be predicted, that spoons can be bent by the power of the mind, or that health problems can be diagnosed by "remote viewing." Judging by the contents of her book *Phenomenon*, Browne believed all this, and more. Self-levitation, psychokinesis, and spirits are all real. So is reincarnation. Sylvia knows. You see, as a "certified master hypnotist," she helped many people overcome burdens imposed by past lives. Such as the young boy who was inexplicably panic-stricken every time his mother prepared to take a shower. Sylvia, through hypnosis, discovered that in a past life he had seen his mother die in the Auschwitz gas chambers, which were disguised to look like showers.

Did Browne really believe all this? Maybe. If she was delusional. But there's another possibility. Perhaps it was all a clever money-making scheme. People apparently waited for weeks to shell out $750 for a thirty-minute psychic reading with this icon of acumen on the telephone. If that sounds a little steep, Christopher, Sylvia's son, was a bargain at $500 a session. It seems he inherited his mother's talents, including the one of fleecing the public. Chris was the only other psychic Sylvia trusted and the only one, other than herself, she ever recommended. A nice little family-run business.

Of course whether I believe in Browne's abilities has nothing to do with whether she had them. That can only be determined by evidence. And that evidence is pretty weak (if we're being generous). For example, back in 1992, Sylvia was unable to foresee that she and her husband would be indicted for investment fraud. I imagine this was not one of the criminal cases she claimed to have collaborated on with the police. As far as other cases go, there is no evidence that she ever played a useful role. She did, however, provide plenty of false leads.

Holly Krewson, a missing twenty-three-year-old, would be found working as an exotic dancer in Hollywood, Sylvia stated on *The Montel Williams Show*. Actually, as was later discovered, Holly had already been dead for six years when the prediction was made. Six-year-old Opal Jo Jennings was abducted from her grandparents' home in 1999. She was alive, Sylvia declared, but had been forced into prostitution and taken to Japan. Sylvia even knew where. A town that sounds like "Kukouro." Geography, like biochemistry, apparently was not her forte. There is no such town. But there was no point in searching for Opal in Japan; she had been murdered by her abductor near the Jenningses' home within a short time of her abduction.

Browne's predictions on other issues also left a lot to be desired. Michael Jackson never went to jail as she had prophesized, but Martha Stewart, who she said would not, did. Elizabeth Taylor was not a big fan, I suspect, since Sylvia repeatedly predicted her death. Needless to say, neither Sylvia nor any other psychic predicted 9/11 or the COVID pandemic, the most impactive events in recent history. Browne did, however, later say that she had terrible dreams the first week in September, but could not figure out why. But as she said, she cannot be expected to be perfect. But how about being a little less imperfect?

Sylvia Browne was repeatedly challenged by James Randi to prove her psychic abilities under controlled conditions. A million dollars was hers if she could do so. Three times Sylvia agreed on network television to be tested. But when actually offered the opportunity, she declined. She claimed Randi didn't really have the money. Well, if she were such a great psychic, she would know that the money was there, locked in an escrow account. Of course Sylvia wasn't really interested in being tested; it is far easier to just go on blathering about the future unopposed. Aliens were to show themselves in 2010 (the same ones who helped build the pyramids), the lost city of Atlantis will reappear in 2023, and by 2100 people will be able to simply walk out of their bodies upon death. Sylvia passed away in 2013, and I guess missed the chance of walking out of her body, whatever that may mean.

THE WATER REVITALIZER

If you live in a major city you probably have some worries about crime in the streets. But what about crime under the streets? Did you know that murder is being perpetrated there all the time, right under our noses? Water pipes are killing our water! I wasn't aware of this criminal activity until I was informed of it by the purveyors of the Original Danish Water Revitalizer, who are dedicated to bringing the crime wave to an end.

What is the problem? Modern water technology. It seems we force water to flow through straight and narrow pipes "with no freedom to follow its innate desire to move in spirals and swirls." We even expose the water to "deadly 90 degree turns." This, I'm told, is a particularly dastardly thing to do. As the unfortunate water molecules smash into the pipe, the bond angle between the two hydrogen-oxygen bonds is reduced, and if it reaches 101 degrees, the water "dies." And dead water of course has no energy to fight off bacteria and cannot service our body properly. So, given that our body is composed of roughly 70 percent water, I guess we should not be surprised that there is so much illness out there.

Ah, but there is a solution. The Original Water Revitalizer will "give the water a double helix spiral which creates a vortex energy field and restructures the water at the molecular level." It is said to restore the water's energy the same way nature does. Apparently nature accomplishes this task by having water flow through winding rivers. Needless to say, the Revitalizer also "removes from the water's memory the electromagnetic frequencies of any pollutants which are just as harmful as the pollutants themselves." All of this is accomplished without any parts that wear out and without any filters that need replacing. Surely a bargain for $298!

And what do you get for your investment? Nothing more than a piece of curved stainless steel pipe that can be attached to your faucet! All those dead water molecules, the victims of the straight and narrow pipes, are now resurrected as they cruise through the life-giving curve. The theological consequences alone are staggering! So are the testimonials offered by people who drink, wash, and shower in "Living Water." They feel more energetic, use less detergent, and no longer have to cope with the taste of chlorine. There must be some novel chemistry here because my chemical education has not prepared me to explain how chlorine is removed by having the water flow through a curve.

So you think I'm making this up? Well, I'm not. I don't have that much imagination. And it really does take a great deal of imagination to assemble as much preposterous pseudoscientific gobbledygook as we see in the advertisements for the various types of water revitalizers. The most scary aspect of this little story is that obviously there are people buying into this scam. They even think they are helping the environment because they are told that the effect of the revitalizer is so strong that when the water is transported back to nature, it cleans the rivers, lakes, and oceans. Instead of attaching curved pipes to our faucets, we need to educate people about science so that they are equipped to straighten out the crooked purveyors of such nonsense.

Battling nonsense, though, is a tough task. Whenever you think you have seen the ultimate in absurdity, something else comes along and reaches even loftier heights. The Centre for Implosion Research in England is also out to publicize water's confrontation with straight motion. Their remedy is not just a curved pipe, but a "vortex." "Natural water forms whirls and eddies, and given the freedom, will form a vortex, as in going down the drain." Vortexes create great energy, we are told. Anyone with doubts is counseled to take a look at what a tornado can do. Life lies in vortexes, we're informed, and are referred to the spiral structure of DNA as proof. It seems it is easy to infuse vortex energy into water. No, you don't have to buy a vortex-shaped pipe to run your water through. You do, however, have to purchase the Vortex Energiser, which is a spiraling copper tube that contains "imploded" water. You just place this next to your water pipe, or near any water that has to be brought back from the dead. Such as the water in fruits and vegetables that have been treated with "chemicals." Just put the Vortex into the fridge and it will solve this problem. We are never told what imploded water is, just that it is the opposite of "exploded" water, which I assume those of us unwilling to invest in the Vortex Energiser are drinking with great risk to our health. Maybe that's why I feel like exploding when I read about imploding water. The Energiser, filled with a lifetime supply of imploded water, goes for about $250.

Now after all this balderdash about water pipes, let me mention a real concern. Prior to about 1970, water pipes were made of cast iron or ductile iron without any internal protective liner against corrosion. After that date, pipes were protected, often by a thin layer of cement on the inside. The unprotected pipes have corroded over the years and have built up scale and rust deposits, often restricting water flow and reducing pressure. In some areas there is concern that in case of fire there would be inadequate water pressure. Water that flows through corroded old pipes comes out rusty, but that isn't the biggest concern. If bacteria get into the water, they can take up residence in the irregular crevices on the corroded surface and multiply, as apparently was the case in Walkerton, Ontario, where over 2,000 people got sick and six died from E. coli contamination. That's one of the reasons why sometimes in communities with old pipes residents are urged to boil their water. Pipes can be rehabilitated by lining them on the inside with cement or epoxy. In many cities in the U.K., the U.S., and Canada, this is being done systematically. Toronto does some 170 kilometers of cleaning and lining every year. In Quebec, it's only a few kilometers a year. Relining water pipes is an excellent investment that will pay dividends in the future. So it seems the Revitalizer people are right about one thing. Water pipes may be troublemakers. But the solution to the problem does not lie in their crooked advice.

WHERE'S THE AURA, ASKS YOUNG EMILY ROSA

Rarely does a report published in the *Journal of the American Medical Association* generate as much publicity as did the one entitled "A Close Look at Therapeutic Touch." Of course, rarely is one of the authors of a paper in one of the world's leading scientific publications an eleven-year-old girl. And rarely is a grade four science fair project the subject of a research article.

How did this account get into the prestigious *Journal*? The same way that any other scientific report makes it into its pages. It passed the scrutiny of expert reviewers. Young Emily Rosa, with help from her nurse mother, described how she had designed a simple experiment to test the claim of therapeutic touch (TT) practitioners to be able to detect the "energy aura" that surrounds a human body. Proponents say they can feel the energy field when they move their hands above a patient's body and that they can even reconfigure it to correct "imbalances." Conditions they claim to be able to help range from arthritis to Alzheimer's disease.

Emily tested twenty-one practitioners in a straightforward manner. They were asked to put both hands through a screen that hid them from the young scientist. She then hovered her hand above one of the subject's hands at a distance that the TT practitioners had previously agreed was ideal for sensing the energy field. All they had to then do was report which one of their hands was sensing Emily's presence. The results were no better than what one would expect by random chance alone.

Not surprisingly, this report sent the proponents of therapeutic touch into mental gyrations. Dolores Krieger, a professor of nursing at New York University, who had pioneered TT in 1973, came out with guns blazing. The study was invalid, she said, because one doesn't just go into a room and perform TT, and furthermore the healer's hands have to be moving all the time to detect the "aura." But in this case, the TT practitioners had agreed before the experiment that the conditions were in fact acceptable and that they would be able to feel the field generated by Emily's hand.

Professor Krieger's objections are understandable since she parlayed therapeutic touch into a huge business. There are close to 50,000 practitioners in North America and TT is taught in many universities. Hospitals often hire therapists at rates of $70 an hour to balance energy fields around patients. Krieger claims this is well justified and points to the "hundreds" of supporting studies in the scientific literature. When one actually looks up these studies, however, they fall far short of proof. Most of them can be far more readily criticized than Emily Rosa's science fair project.

While Emily's study showed that TT does not work the way its proponents claim, it certainly did not invalidate its use. There is no doubt that many people find relief when they believe that their energy fields are being manipulated. When a patient feels better, the why becomes of secondary importance. If therapeutic touch works through the placebo effect, then so be it.

Dolores Krieger of course believes that there is more to TT than the power of suggestion. She first got interested in the subject in 1971 when she was part of a research team that studied the supposed healing abilities of the remarkable Oskar Estebany, a former Hungarian cavalry colonel. Estebany claimed to heal horses and people just by the laying on of his hands. There were enough testimonials to prompt a study by Bernard Grad at McGill University in the 1960s in which the healer was apparently able to control the rate at which goiters grew in rats deprived of iodine by placing his hands around the cage for fifteen minutes twice a day. In Krieger's study, Estebany appeared to be able to affect patients' hemoglobin levels. This was enough to sell Krieger on the idea of healing by hand motion and she soon discovered that she could unblock patients' congested energy (even though this is unmeasurable by any known device) and could even infuse them with her own. To her credit, Krieger only works together with medical doctors and claims only to be able to cause relaxation and pain relief. There is no talk of miracle cures.

Dolores Krieger of course was not the first to suggest that the human body comes equipped with some sort of spiritual energy that governs health. Indeed, this is one of the most ancient concepts in what today we may call alternative medicine. The Chinese have long believed in the mysterious life force called qi (pronounced chi) that travels through the body's "meridians" and which can become imbalanced, resulting in disease. Correction of this imbalance by acupuncture, patterned breathing, or diet then affords relief. The age-old practice of Ayurvedic medicine in India embodies similar ideas. The human body is seen to be made up of energy elements called doshas, which operate through body channels, and their proper flow is critical to health. Neither of these belief systems has any basis in anatomy and arose because both in China and India dissections were forbidden, making the physical workings of the body a matter of great mystery. Healing therefore had to be based on metaphysical beliefs.

These beliefs have of course often proven to be potent and have withstood the test of time. Their success is probably due to a combination of the power of the imagination and the fact that many illnesses are self-limiting or psychosomatic. Any alternate explanation would force scientists to swallow pretty hard. How, if not for the power of suggestion, could we explain the healing abilities of George Chapman, an uneducated Englishman with no medical knowledge. He claimed to have been contacted by the spirit of a deceased physician who taught him to go into a trance and operate on the spirit body of a patient with invisible surgical instruments. Not only were there plenty of testimonials, but patients even said they felt the twinge of the scalpel and the drawing together of the spiritual flesh after the operation.

Wilhelm Reich, a psychoanalyst who trained under Freud, did not look to spirits to reveal details about the body's health-governing energies. He looked to outer space. Reich believed that he was the product of a relationship between an alien and an Earth woman and that this unusual background allowed him to realize that not only the body, but the universe, was governed by "orgone energy." He derived the word from orgasm, which he thought was the ultimate expression of this form of energy. Illness was due to a deficiency of orgone, which he could remedy by placing the patient in an "orgone accumulator," a box about the size of a phone booth with no mechanical or electrical components. Testimonials abounded.

Not all orgone was good. Reich said that some UFOs are propelled by orgone motors (one wonders what the aliens are doing inside and for how long to generate the orgone) and that Deadly Orgone Energy accumulates in the atmosphere, causing disease on Earth. Of course, a device called a Deadly Orgone Buster was available to rid us of this scourge.

Believe it or not, the orgone promoters are still with us. One of them sells an orgone generator for the home that is even capable of remote energizing as long as a "transfer disk" is carried in the pocket. They offer proof of the device's efficacy on the Internet. Just download a "transfer diagram" and hold your hand three inches above it to experience a sensation of warmth or a gentle cool breeze, they say. I tried it. I felt no orgone. The only thing I felt was silly.

INFOMERCIALS PROVIDE SLANTED SCIENCE

Sometimes when I have a touch of insomnia I'll turn on the TV. Cruising through the channels some years ago, I came across Larry King, sporting his trademark suspenders, interviewing a guest on a program called *Larry King Special Report*. I knew Larry had left CNN and was hosting a couple of interview shows on the RT (Russia Today) and Hulu cable channels but I don't get these, so what was I watching? It soon became apparent that this was not a legitimate interview show but rather an infomercial. King was shilling for a dietary supplement, Omega XL, going on about how it provides miraculous joint pain relief. His "guest" was Dr. Sharon McQuillan, waxing poetic about how she recommends Omega XL to all her patients to "help protect their hearts, preserve their heart and vascular health." Larry, who has a history of heart disease, asked McQuillan how Omega XL can reduce the risk of heart attacks. She answered that "thirty years of studies have shown the benefits of omega-3s." That is true, but totally misleading since none of those studies used this product!

Some studies have indeed shown that eating foods rich in omega-3 fatty acids may lower the risk of death from heart attack. And trials with supplements containing DHA and EPA, the two major omega-3 fats found in fish, have suggested a benefit for people who have previously suffered a heart attack. For example, in a placebo-controlled trial, patients taking an omega-3 supplement were 6 percent less likely to suffer a decline in heart function as determined by magnetic resonance imaging (MRI) than those taking a placebo. That's not a very

Uncertain about what to do? Just pick up the book on Bach Flower Remedies and discover how scleranthus extract is the answer to vacillation, indecision, and uncertainty.

What is the common feature of all these therapeutic approaches? They all come oozing piles of anecdotal evidence. Breast cancer victims describe how breast lumps disappeared after going on an organic juice diet, and patients with indigestion reveal how they improved after ridding their intestines of "parasites" with some miraculous herbal concoction. It all sounds great. The only problem is insufficient evidence. There are no controlled studies to back up the claims, no follow-up investigations to see whether the reported "cures" were maintained. Of course, the lack of controlled studies does not mean that a particular treatment does not work. After all, many medical discoveries start with anecdotal evidence. Someone makes an observation, which at first may even appear to be outrageous. Like noting that eating limes prevents scurvy. James Lind, a Scottish physician, was ridiculed in 1754 when he suggested that sailors be given citrus fruits on long sea voyages to prevent this dreaded disease. But soon his observation was confirmed. Sailors who maintained the usual diet of dried bread and salted meat got scurvy, whereas those who supplemented the meager fare with limes did not. Anecdotal evidence changed into scientific fact.

It is certainly possible that some of the remedies and treatment regimens that vie for shelf space in the health food stores will cross the bridge from conjecture to scientific fact. But until that happens, they remain in the category that we have come to refer to as alternative therapies. This does not necessarily mean that they are ineffective, only that they are untested or unproven. What is a proven, however, is that people are flocking to these therapies. Why? Because modern scientific medicine cannot provide cures for all ailments and in many cases physicians are perceived as beleaguered, uncaring, and unaccepting of new ideas. The alternative practitioners are usually charismatic, spend a great deal of time with the patient, and imply significant difference. And they were taking four grams a day! Omega XL contains 6.3 mg of EPA and 4.9 mg of DHA, roughly 1/400th the dose that showed a minor benefit in the study! In other words, there is no basis to suggest that the EPA and DHA in this supplement can help protect the heart.

Contrary to the image projected by the "interview," Omega XL is not a DHA/EPA supplement. It is an extract of the "green-lipped mussel" found in New Zealand and is a complex mix of many compounds. There is some evidence for an anti-inflammatory effect and for possible benefit in arthritis and perhaps even in asthma. But Dr. McQuillan's enthusiastic recommendation for protecting the heart is not supported by evidence. As is revealed in the credits after the show, McQuillan, who is a GP specializing in "integrative, regenerative and aesthetic medicine," was paid for her appearance.

Of course Larry King is not the only celebrity who has lent his or her name to promoting a product. In fact, back in 2000 when Larry was still on CNN, he featured Olympic champions Dorothy Hamill and Caitlyn Jenner as guests. *Larry King Live* was certainly not an infomercial; it was one of the most respected and most watched interview shows on TV. Both guests were there to talk about the painkilling drug Vioxx. "My doctor prescribed Vioxx for me, and it's as if I've been given a new life," Hamill told King. "It's just, it's been amazing. I feel twenty years younger." Jenner won the decathlon at the 1976 Montreal Olympics but subsequently had knee surgeries and shoulder problems that had left her with pain that resolved with Vioxx. Both athletes were paid by Merck, the drug's manufacturer, something that was made clear on the program.

Neither Jenner nor Hamill could have known at the time that Merck was already investigating an apparent increase in heart attacks among people taking Vioxx. Four years later the drug would be recalled for that very reason, precipitating some 35,000 lawsuits and payments of over $4 billion by Merck to the plaintiffs. Were there people who were prompted by Hamill and Jenner's appearance to ask their doctors to prescribe Vioxx? Undoubtedly. Did some suffer adverse consequences? Who knows? But a supposedly objective interview show is no place for paid celebrity product endorsers.

There is also the thorny question of television ads for prescription drugs featuring celebrities. Legendary golfer Arnold Palmer and basketball star Chris Bosh had health problems that required treatment with an anticoagulant, and in a TV ad they sang the praises of Xarelto (rivaroxaban), an effective oral "blood thinner." Although this is a prescription drug, and the side effects and risks are outlined, the ad still indirectly suggested taking medical advice from a celebrity with no relevant expertise. Kim Kardashian, who is famous for being famous, once promoted Diclegis, a morning sickness relief pill for pregnant women. The efficacy of the drug is questionable, and Kim's lighthearted tome on Instagram — "OMG have you heard about this?" — prompted the FDA to censure her for not including risk information or limitations for the use of the drug, something she subsequently corrected. Was there any damage done? Who knows?

Recall the lyrics from "If I Were a Rich Man" from *Fiddler on the Roof*: "And it won't make one bit of difference if I answer right or wrong, when you're rich they think you really know." That seems to embody the philosophy of advertisers who hire celebrities. But no matter how rich you are, you can't buy knowledge.

DIAGNOSING PATHOLOGICAL SCIENCE

Back in 1953, Nobel Prize–winning chemist Irving Langmuir coined the expression "pathological science" to describe a process by which a scientist seems to follow the scientific method but unconsciously strays in favor of wishful thinking. Pathological science is distinct from fraud in that there is no intent to deceive. It is essentially faulty science promoted by people who are somehow blind to the evidence against their pet ideas. The most frequent use of the term over the last couple of decades has been in connection with "cold fusion," a phenomenon first proposed in 1989 by electrochemists Stanley Pons and Martin Fleischmann.

The two highly regarded researchers stunned the scientific community by calling a press conference to announce that they had detected the fusion of deuterium nuclei under simple laboratory conditions. They claimed to have evidence of release of energy that could not be explained otherwise. A wave of excitement spread around the globe with optimistic musings about the process that could be the answer to our energy problems. The euphoria quickly waned when other research groups were unable to reproduce the experiment. A general consensus soon emerged that Pons and Fleischmann had noted some anomalous phenomenon but had misinterpreted their findings. It was wishful thinking, not experimental evidence, that had produced cold fusion. Actually, while the names of Pons and Fleischmann are most commonly associated with cold fusion, they were not the first to claim that such a process can occur under mild conditions.

Some thirty years before Pons and Fleischmann's press conference, French chemist Louis Kervran introduced his theory of "biological transmutation," claiming that in living systems atoms of one element can combine with those of another to give rise to a third element. Potassium, Kervran suggested, can under the right conditions combine with hydrogen to form calcium. This was a stunning claim, an apparent realization of the classic alchemical quest to transmute one element into another. Could this be?

The identity of an element is determined by the number of protons in its nucleus. Potassium has nineteen protons and hydrogen has one. If in the body these somehow combined to form one nucleus of twenty protons, we would indeed have calcium. We would also have an event that defies everything we know about chemistry and physics. Elements don't combine to form new elements except in the case of nuclear fusion reactions, which require a tremendous input of energy only achievable at temperatures of millions of degrees. Such conditions are met in our sun, where hydrogen nuclei combine to form helium, accompanied by the release of vast amounts of energy. But outside of such extreme conditions, chemical reactions cannot create or destroy atoms, they can only rearrange them to form novel molecules.

Kervran had originally been intrigued by a question raised by French chemist Louis Nicolas Vauquelin in the late 1800s. How could hens manage to produce eggshells, which are composed of calcium carbonate, in spite of being fed a diet of oats, known to be very low in calcium? Kervran concluded that potassium, which is plentiful in oats, must combine with hydrogen to produce calcium. He dismissed the problem of the immense energy requirement for such a process by suggesting the existence of a "low-energy transmutation." This would become known as the Kervran effect. Kervran also claimed that a crayfish placed in a basin of sea water with all calcium removed still managed to make a shell. Again, he suggested, potassium must have been converted into calcium.

According to Kervran, even the strange case of industrial carbon monoxide poisoning when no carbon monoxide had actually been inhaled was to be explained through transmutation. Nitrogen gas, which makes up about 80 percent of air, is composed of two nitrogen atoms joined together. Each nitrogen atom has seven protons in its nucleus, and if a proton from one nitrogen atom were to be transferred to the other, two novel nuclei, one with six and the other with eight protons, would be formed. In other words, we would now have an atom of carbon and one of oxygen, thereby explaining the formation of carbon monoxide. But chemistry doesn't work by such simple arithmetic. Such a transfer of proton from one nucleus to another has never been observed.

In spite of the scientific implausibility of biological transmutation, Kervran's theory was not dismissed out of hand by all. Italian researchers carried out a carefully controlled study of oats under a variety of conditions, analyzing for calcium, potassium, and magnesium. There was no evidence of any kind of transmutation. What then about the egg-laying conundrum? Actually, there is no conundrum. If there isn't enough calcium in the chickens' diet, the birds will mobilize calcium from their bones. Furthermore, oats are not devoid of calcium; the early analyses back in the nineteenth century were faulty. Very simply, the calcium needed for eggshell formation must somehow be provided in the diet. It doesn't come from any sort of transmutation. In modern egg-laying facilities the diet of the hens is often supplemented with crushed oyster shells, cuttlefish, or crushed limestone to ensure adequate calcium intake. Sometimes even eggshells themselves are recycled in feed.

Kervran's idea about carbon monoxide production in the body from nitrogen was also wrong. There is no mystery about finding carbon monoxide in the blood of people who never inhaled any. It forms naturally in the body when an enzyme called heme oxygenase reacts with heme, a breakdown product of hemoglobin. The bottom line is that the Kervran effect doesn't exist. The French chemist simply came to the wrong conclusion based on some faulty observations.

It is curious that in spite of being a competent and respected scientist, indeed an expert on radiation poisoning, he was willing to propose a theory that flew in the face of established science. No wonder he was awarded the 1993 Ig Nobel Prize for physics in recognition of his conclusion that calcium in chicken eggs can be created by some sort of biological transmutation. Were such a fusion process to occur in a chicken, the energy released would turn the bird into an atom bomb.

The Ig Nobels are awarded annually to "honour achievements that first make people laugh and then make them think." Kervran's "biological transmutation" was well worthy of the award. If there were awards for pathological science, Kervran would be a highly ranked candidate. Of course there would be many others jostling for spots.

PUTTING PIMAT'S HEALTH CLAIMS TO BED

Over the years a variety of exuberant and hopeful marketers have trooped into either my office or living room with all sorts of newfangled cleaning products, dietary supplements, water filters, air purifiers, and cure-alls. I've listened to countless sales pitches for smell-absorbing seat cushions, gopher deterrents, healing pendulums, negative ion generators, energizing bracelets, radiation neutralizing pendants, magnetic shoe inserts, and even crocodile traps. But perhaps the most memorable health aid I've ever been asked to examine was Pimat. What, you are asking, is Pimat? Hang on, it's going to be a bumpy ride.

My story begins with a call from two young ladies who asked if I would be interested in taking a look at a "revolutionary, cutting-edge, breakthrough product" they had come across on a recent trip to Poland. It was the answer to sleep disorders, headaches, depression, back pain, asthma, arthritis, impotence, and fertility problems. I mentioned that I wasn't in the market for the latter two, but that I did sometimes have insomnia thinking about the numerous scam products promoted by con artists. They agreed that there were a lot of snake oil salesmen out there, but they had found something that really works. They knew this because they had tried it themselves and felt energized and mentally clarified! Sounds great, I said, more energy and mental clarity are always welcome. Bring on the Pimat! And they did.

The wondrous discovery turned out to be a piece of cloth, about 18 centimeters square, adorned with an unsymmetrical pattern of ten red dots. I stared in amazement at the fabric that was to be the answer to so many of our health problems.

"Pimat," I learned, stands for "pyramid mat" and "reproduces, in two-dimensional form, the healing power of the pyramids." The pattern of red dots "rebalances and restores our energy fields, our aura, because it generates an entire spectrum of 'radiesthetic colours' essential for the body to maintain a healthy condition." Ohhh . . . kay. And how do you avail yourself of the powers of Pimat? Simple. You put it under your sheet and you sleep on it.

What scientific evidence is there, I sheepishly asked, that the health of our aura is being restored by this wondrous piece of spotted fabric? A hand quickly went into the bag and emerged with a 184-page treatise on this magnificent "regenerator of stamina." On the very first page I was introduced to the inventor, Ryszard Olszak, from whom I learned that the action of this product is based on the energetic effect of its geometrical configuration, or "neoenergy." He does admit that "in spite of much research, the nature of this phenomenon could not be explained in terms of physics, and the energy emitted has not yet been measured by any conventional methods." Yet, he knows it exists. To characterize neoenergy, I was informed, "radiesthetic colours are used." What are these? Let's go to the horse's (or perhaps in this case, ass's) mouth.

"The aura is composed of all the colours of the spectrum, characterized by different wavelengths corresponding to colours, known as 'radiesthetic colours.' If one is ill or tired [off-colour!], there will be colours missing or faded or holes in the aura." It seems that in 1991, Ryszard Olszak discovered, through Radionics, a pattern that generates an energy called neoenergy or the energy of shape. "This produces all the colours of the spectrum and restores any that are missing." There you have it. Convincing rationale as to why that Pimat works.

Not convinced? The Pimat document is full of diagrams, Kirlian photographs, charts, and of course testimonials galore. Pains vanish, infertile women conceive babies, bedwetting disappears (surely of comfort to Pimat), and people experience more lucid dreams because "pyramids generate fractal energy fields." The only warning is that the efficacy of Pimat may be reduced if the bed is against a wall that has a fridge or TV on the other side. In such a situation we are urged to "contact your dowser or feng shui consultant." Needless to say, Pimat has no side effects. That, unquestionably, is a true statement.

While Pimat can perform numerous miracles, I was disappointed to learn that it may not be able to neutralize "geopathic stress." That, you should know, occurs when the Earth's electromagnetic field becomes distorted. I learned that "the Earth resonates with an electromagnetic frequency of approximately 7.83 hertz (Schumann resonance), which falls within the range of (alpha) human brainwaves. Underground streams, sewers, water pipes, electricity, tunnels and underground railways, mineral formations and geological faults distort the natural resonance of the Earth thus creating geopathic stress." Pimat's powers are apparently tested to the extreme when dealing with geopathic stress, but luckily the HELIOS 3 Geopathic Stress & EMF Home and Office Harmoniser, which can be plugged into any electrical socket, can ride to the rescue.

By this time my head was spinning, perhaps from all the "radiesthetic energy" that I had absorbed from fondling Pimat. But I thought that there was perhaps a teaching opportunity here. I suggested to the ladies that we run a little test. We would enlist a number of subjects who would sleep either on Pimat or on a piece of the same fabric without the red dots. Both the Pimat and the control fabric would be sewn into coded cotton envelopes so nobody would know who was sleeping on what. I had been assured by the manufacturer that the cotton envelopes would not interfere with the therapeutic effect. Indeed, given that the product is supposed to work its magic when placed under the bedsheet, they could hardly claim that the wondrous neoenergy could not pass through a cotton envelope.

I managed to enlist about eighty senior subjects who, before even being told what the experiment would be, were asked to fill out questionnaires about their health and energy status every morning for two weeks. For the following four weeks, half the subjects slept on Pimat and half on the placebo fabric. They then switched samples and carried on for another four weeks, filling out questionnaires every morning.

As I had expected, the presence of red dots made absolutely no difference. But there was a significant difference between the first two weeks, when there was no intervention of any sort, and the next eight weeks. Whether they slept on an "active" Pimat or on the placebo version, people claimed that they had slept better and woke up with more energy. Some even claimed that they had been able to perform conjugal activities during the night with greater vigor.

To their credit, the two ladies recognized that they had witnessed an example of the placebo effect and decided not to market Pimat as had been their intention. Others may not be so insightful. One Pimat devotee proclaims that her alternative doctor confirmed that her immune system was now at 80 percent. I wonder what breakthrough technology she uses to make such a measurement. In any case she was so impressed that she ordered twenty-two Pimats for her patients, at about $20 each. Of course couples sleeping together need two Pimats.

Although our study of Pimat was not rigorous enough for publication, the demonstration to my new acquaintances of the importance of carrying out controlled trials did leave me with a degree of satisfaction. I was left with something else as well. A number of Pimats. And I have since discovered that they really do work. At cleaning computer screens.

CHEMICAL-FREE IS NOT A GOOD DEAL

A newspaper report about salmonella contamination found in hydrolyzed vegetable protein, a common flavoring agent, began thus: "It sounds more like a chemical than a food ingredient . . ." Well, if a food ingredient isn't a chemical, what is it? Of course it's a chemical. Everything in the world is made up of chemicals, which are nothing other than the building blocks of all matter. The oxygen we breathe, the water we drink, the sugar we eat are all chemicals, as are the medications we swallow, the cosmetics we apply, and the pesticides we spray. But somehow "chemical" has become a dirty word, synonymous with "toxin," and "chemical-free" is now a popular, albeit ridiculous, advertising slogan. Chemicals are not good or bad, dangerous or safe. They don't make decisions. We do. And those decisions should be based on science, not emotion.

A chemical's properties are determined by its molecular composition and structure, not by its ancestry. Whether the molecule was made by nature in a plant or by a chemist in a lab is irrelevant. It is what research has revealed about its properties that matters. And there is a stunning amount of such research. In 2021, the American Chemical Society's Chemical Abstracts Service registered the 144 millionth known compound! These include both natural and synthetic substances that have been described in the scientific literature or in patents. The landmark fifty millionth compound was registered just about a decade earlier, demonstrating the remarkable speed with which chemistry advances. That compound had a Canadian connection, having been developed by Montreal's Chlorion Pharma, as a potential treatment for neuropathic pain. Chemical Abstracts lists it as (5Z)-5-[(5-Fluoro-2-hydroxyphenyl)methylene]-2-(4-methyl-1-piperazinyl)-4(5H)-thiazolone!

Some eyes are probably rolling now. Who would want to inflict a chemical with such an unpronounceable name on their body? Well, the number of letters in a name has no more to do with a substance's properties than does its "natural" or "synthetic" origin. Obviously, when dealing with over a 140 million known compounds, each requiring a unique name, complex terminology has to enter the picture. Chemists are thankful for the systematic nomenclature that has been worked out, but as far as the public goes, complex chemical names are frightening and tend to conjure up images of doom. Some marketers attempt to capitalize on this fear by advertising "chemical-free" products.

And so we have "chemical-free" cosmetics, cleaning agents, and, believe it or not, books about "chemical-free kids." The message is that chemical-free means safer, healthier, greener. Given that it is a nonsensical term, what are these products all about? Mostly, "chemical-free" refers to being free of synthetic chemicals. This of course insinuates that synthetic chemicals are more problematic than natural ones, an implication that is not valid. Take, for example, the case of "chemical-free" sunblocks. These are often based on titanium dioxide, a naturally occurring mineral. Certain formulations of titanium dioxide have raised safety concerns and the chemical has even been classified as a carcinogen when inhaled. It seems safe enough in sunblocks, but the designation of such products as "chemical-free" is sheer nonsense.

Mainstream food producers are also trying to capitalize on the anti-chemical fervor. McCain Foods, for example, ran a campaign that touted only "real ingredients" in its pizzas. What does that mean? Were they using imaginary ingredients before? Or perhaps fake ones? Plaster of Paris instead of flour? Play Dough instead of cheese? Here's what McCain says: "It's all about the ingredients. And good food, frozen or not, starts with real ingredients. We know that when you look at an ingredient list you want to see familiar ingredients, not ingredients you can't pronounce." Makes me want to scream some words that can be pronounced easily.

According to the ads, McCain aimed to remove "unfamiliar ingredients." Specifically mentioned were sodium stearoyl lactylate and sodium ascorbate. Why remove these? There is absolutely no scientific reason; it is all a question of marketing. Both are approved food additives and have undergone rigorous testing. Sodium stearoyl lactylate is an emulsifier used in baked goods, like pizza dough. It disperses the fats in the dough, allowing less fat to be used while softening the dough's texture at the same time. Since it is made from lactic acid, found in milk, and stearic acid, found in beef tallow, you could even call it "natural." Sodium ascorbate is just the sodium salt of vitamin C, and is used as an antioxidant to prevent fat from going rancid. These additives actually make for a better dough. Removing them just caters to the current wave of chemophobia.

McCain also made a big deal out of using only vine-ripened tomatoes. A noble endeavor. Vine-ripened tomatoes certainly do taste better. And the riper the tomato, the more natural ascorbate it contains. So while the company sings the praises of taking out ascorbate on one hand, it actually increases the amount of the same chemical with the other. Of course it's all silliness because there is no problem with sodium ascorbate or stearoyl lactylate in the first place. And curiously, while McCain's was touting the elimination of ascorbate from its pizza dough, it happily promoted the presence of vitamin C, a less daunting term for ascorbate, in its potatoes!

We live in a chemical world with a novel substance being isolated or synthesized roughly every 2.6 seconds. Rather than representing a cause for worry, this just shows the amazing progress of science. Most of these new chemicals will never become anything other than listings in Chemical Abstracts, but some will become key ingredients in new drugs, fabrics, plastics, electronics, and myriad other items, which certainly won't be "chemical-free." But if you insist on buying a truly chemical-free product, remember that you won't be getting a good deal. You'll be buying something that contains nothing.

JAMIE OLIVER COOKS UP SOME NONSENSE

Celebrity chefs are cooking up a storm these days. But at least one has also brewed up a controversy. Jamie Oliver opened up quite a can of worms when he planned to improve the food served in Los Angeles schools. Actually, worms would probably be an improvement over some of the fatty processed food served, but after initially agreeing to the filming of Jamie's popular *Food Revolution* program inside its schools, the Los Angeles school board decided that while the chef's ideas for improved nutrition would be welcome, his cameras would not be. Apparently they had heard that a similar venture by Jamie in Huntington, West Virginia, left a bad taste is some administrators' mouths.

Let me state for the record that I like Jamie Oliver. I value his attempts to improve students' health through his "Food Revolution." But I'm revolted by his "science," or I should say his lack of it. No big surprise here, I guess, given Jamie left school at the age of sixteen to pursue his culinary interests at Westminster College, where I suspect chemistry was not a highlight of the curriculum. But the British chef's sketchy grasp of science has been no impediment to his success. Jamie Oliver has become a true phenomenon, writing books, opening up restaurants, selling kitchen gadgets, performing in theaters, cooking for dignitaries, and of course starring in a variety of TV shows.

Unlike many self-proclaimed nutritional gurus, Jamie is not an extremist. While he favors organic ingredients when possible, he doesn't espouse a vegan diet and is not averse to a hamburger. He just wants that hamburger to be made of proper ground beef, not various meat by-products. And if kids are to have an ice cream sundae, he'd rather "it didn't contain shellac, hair or beaver glands." It is with statements like this that Jamie muddies the waters.

While the Los Angeles school board was reluctant about allowing him to exercise his culinary talents in the kitchen, one school agreed to let Jamie "teach" a science class about food. And these kids were sorely in need of some education along these lines, considering that some thought honey comes from bears and chocolate is pumped from a chocolate lake. Jamie's science class, though, came down to frightening students away from processed foods with a made-for-TV, dramatic, but nonsensical demo.

"Do you know what is in your ice cream sundae?" Jamie asked the class as he prepared to reveal the "truth." Out came a blender and in went a mix of live lac bugs, human hair, and feathers. Surely beaver glands would have been thrown in had they been available. Instead Jamie had to make do with a stuffed beaver forlornly looking on, as if deprived of his anal glands by nasty food chemists looking to improve the flavor of ice cream. What a blend of nonsense!

Let's start with the shellac, a secretion of the lac bug that in a purified form can indeed be applied as a coating on the candy topping that decorates sundaes. But implying that chopped live insects are an ingredient in sundaes is ridiculous. As ridiculous as the theatrics with hair and feathers. Here, the reference is to cookie dough that may be found in some ice creams and is often formulated with L-cysteine, an amino acid that improves texture. This compound can be readily isolated from the mix of amino acids produced by chemically breaking down proteins. Indeed, both hair and feathers are composed of proteins and can serve as the raw materials for the production of L-cysteine. These days cysteine is actually made by a fermentation process, but its origin is really irrelevant. What matters is what the final product is. And L-cysteine is a harmless, approved food additive. Any suggestion that duck feathers are added to ice cream is quackery.

On to the beaver glands. Jamie debuted this piece of puffery on the *Late Show with David Letterman*. David has been sworn at by Cher, instructed on the use of cucumbers in the bedroom by Dr. Ruth Westheimer, and has had to endure exposure to a variety of excreta from his animal guests. But rarely has he expressed the kind of shock we saw when chef Oliver blurted out that cheap strawberry syrup and vanilla ice cream can contain beaver anal glands. Audience reaction mirrored Letterman's, and the episode triggered predictable exchanges on the web expressing outrage about "what they're putting into our food." Some people wondered about just how many beavers sacrificed their lives to produce that liter of ice cream in the fridge. The answer to that is none. But it is true that when beavers are trapped for their pelt, admittedly not a pleasant thought, two small glands near their anus that produce a territorial marker called castoreum are removed, and their contents extracted with alcohol for commercial use. A few parts per million of purified castoreum may be one of the ingredients included under "natural flavor" on a jam or ice cream label. But that's a long way from mixing beaver glands into ice cream.

So we have an interesting philosophical question here. Does the end justify the means? Is it acceptable to improve people's diet by evoking the yuck factor through false information? Jamie Oliver strives to feed people a proper diet, which is great, but he is also putting their brains on a diet devoid of science. In my view, teaching the wrong thing is never right.

I wonder what Jamie thinks about drinking the mammary gland extract of a cow? Or eating the ovum of a chicken? Has he ever had any escargots? It's also interesting to note that while he was terrifying Letterman with prospects of beaver glands in his ice cream, Jamie was cooking up mussels. Have you ever seen what raw mussels look like? Beaver glands appear positively charming next to this slimy tissue. If you want to worry about something in ice cream, worry about the high sugar and fat content. As far as the beaver glands go, well, obsessing about them is, let us say, anal.

A TREATMENT FOR CHEMOPHOBIA

"Hey, aren't you somebody?" the teenager queried as I got into the elevator. While I was pondering an appropriate answer to this deeply philosophical question, his crony spilled the beans: "Yeah, he's that guy who talks about chemistry on TV." This was just the ammunition the philosopher needed. "Oh no, we're locked in an elevator with a scientist," he mocked, before volunteering the information that he got about 2 percent in chemistry in high school, and "that was with cheating."

Sadly, I've heard such comments before. After many a public lecture I've been approached by people who somehow feel the need to unburden their soul and tell me with some sort of perverse pride how they slept through science classes or that chemistry was the only course they ever failed. Little wonder that there is mental chaos about chemicals and that "chemical-free" products are hot sellers.

Chemical absurdity has even made it into the courtroom. The prosecutor in a gang fight trial in California described "a situation very much like nitrogen meeting glycerin; it was guaranteed that there would be an explosion of violence." He probably had some vague recollections about nitroglycerin being a potent explosive. But this substance is not made by combining glycerin with nitrogen. Actually, glycerin meets nitrogen all the time quite peacefully since nitrogen makes up 80 percent of air!

In a more serious vein, cleanup crews dressed in decontamination suits descended on a small town to deal with a toxic emergency caused by a mercury spill. The culprit was not some careless chemical company; the culprit was chemical ignorance! A couple of teenagers found a 20-kilogram batch of pure mercury in an abandoned neon light factory and proceeded to have some great fun with the shimmering substance. They played with it, distributed some to friends, spilled it on the floor at home and at school. As a result, eight homes had to be completely emptied of furnishings and six students ended up in hospital, where they had plenty of time to contemplate the dangers of mercury, dangers they should have learned about in high school chemistry class!

This mercury episode is pretty scary in terms of what it says about science education. But even more chilling is the story of young Nathan Zohner, who won the Greater Idaho Science Fair by getting forty-three of fifty passersby to sign a petition to ban dihydrogen monoxide because it can be fatal if inhaled, is a major component of acid rain, and can be found in the tumors of terminal cancer patients. What was this horrible chemical? Water, of course (H_2O)!

You've guessed by now that the preceding is an appeal for more and better science education at all levels. We are in trouble when a survey among thirteen- and fourteen-year-olds reveals that they view scientists as "nerds and losers who are not accepted by society because they do not want to be." We are in trouble when a magazine advises its readers to drink water frequently because "one third of water is oxygen and drinking it will keep you alert." We are in trouble when it is possible to graduate from high school without ever having had a full course in chemistry, physics, or biology. We are in trouble when a university student describes his professor as "someone who talks in someone else's sleep."

But there are also some positive signs. Science fairs in high schools are mushrooming. Educational institutions now offer programs that emphasize everyday applied science instead of esoteric theory. And we have the USA Science and Engineering Festival in Washington, which exposes thousands to the wonders of science. It is always a thrill for me to accept an invitation to present at the festival because it offers a chance to implant the seeds of science in some fertile minds and to meet some students brimming with ingenuity. Last time I met one who had developed a way to paint a toilet seat with a luminous chemical so that it could be easily located in the dark! I suspect he won't be signing any petitions to ban dihydrogen monoxide.

RAW WATER

There's dumb, dumber, superdumb, and hyperdumb. Then there is dumb that is beyond a suitable qualifier. Promoting "raw water" as a healthy commodity is in that category. This is water that has not been treated in any way; no filtration, no chlorination, no ozonation, no ultraviolet light treatment, none of those nasty technologies that save millions and millions of lives a year.

Depending on the source, raw water may be perfectly safe, or it may lead to a battle with hepatitis, giardiasis, or a host of bacterial diseases. You could get away with nothing more than cramps, vomiting, or explosive diarrhea, but you could also end up in a close and personal relationship with raw underground water. Six feet underground.

Promoting untreated water in the face of everything we know today about waterborne disease is so preposterous that it shouldn't even merit discussion. But the raw water craze can't be ignored because people are wading into the claptrap. Where? Mostly in California, the very state that introduced proposition 65, requiring warnings like "This Disneyland resort contains chemicals known to the state of California to cause cancer and birth defects or other reproductive harm." That's because there might be a rogue vacuum cleaner somewhere with a touch of lead solder. No real risk with that, but apparently California has no worries when it comes to the real risk of untreated water.

Live Water is a brand of raw water for which some folks happily fork out $36.99 for two and a half gallons. It is the brainchild of Christopher Sanborn, a man who rechristened himself as Mukhande Singh, perhaps because that name goes along better with pictures of this longhaired, raw water guru sitting naked in a yoga position, apparently floating above some remote spring.

As he tells the story, one day Mukhande had a revelation. He discovered that "all bottled water is filtered, sterilized and irradiated for cheaper transport and shelf stability similar to how juice and milk products are pasteurized to save costs." Whoa! These measures are not carried out to save costs, but to save lives! He then goes on to say that "unfortunately this destroys five healthy probiotic strains not found in any other food source and without these probiotics we are unable to assimilate all the nutrients found in our food."

"Probiotic" is a popular catchword these days, and for good reason, given that there is accumulating evidence of the role that gut bacteria play in our health. Probiotics are the "good bacteria" that are thought to overwhelm potentially disease-causing varieties. Surprisingly, Mukhande is clever enough to capitalize on this idea. As long as nobody asks for any evidence.

Because of his concerns about the travesties inflicted on bottled or tap water, Mukhande set out on a search for water untainted by the hand of man and found salvation in a spring somewhere in Oregon. "The first time I drank living spring water a surge of energy and peacefulness entered my being; I could never go back to drinking dead water again." And apparently he knows all about dead water. It's "toilet water with birth control drugs in them," he proclaims, the quality of his grammar being comparable to the quality of his science.

By contrast, Mukhande's Live Water contains a host of probiotics, or at least bacteria he calls probiotics. Analysis of Live Water does reveal the presence of various bacteria such as Pseudomonas putida, one of his supposed probiotics. There is no evidence that this bacterium confers any benefit, but I can point out a paper in the medical literature with the title "A lethal case of Pseudomonas putida bacteremia due to soft tissue infection." By mentioning this, I don't mean to imply that Pseudomonas bacteria make Live Water dangerous; just that there is no evidence they make it "healthy."

Trying to put some oomph into the claim that "living spring water is the key to unlocking a perfect micro-biome balance," Live Water promotions refer to a scientific publication entitled "Non-pathogenic microflora of a spring water with regenerative properties." Alluring, until you take the trouble to read the paper. It has nothing to do with drinking spring water! The paper describes "experimental fresh wounds in an animal model showing reduced inflammation when treated with Italian Comano spring water." The authors hypothesize that this may be due to beneficial bacteria found in the water, and suggest that these microbes may even explain why people believe that bathing in spring waters may have some benefit when it comes to human skin ailments. Absolutely nothing to do with drinking Live Water!

Live Water does come from a remote spring, and there is no evidence that this particular raw water harbors disease-causing organisms. But what the whole concept of raw water does harbor is enough detritus to pollute science and possibly compromise health. Anyone for playing Russian roulette with raw water?

"THE PEOPLE'S CHEMIST"

Ill-informed, self-declared experts yelping from atop their soapboxes are a dime a dozen these days. But it is one thing to hear nonsensical rhetoric about the need to avoid substances with unpronounceable names from some scientifically illiterate blogger, and quite another to witness a legitimate chemist urging people to adopt a "chemical-free" lifestyle. Yet those are the exact words used by Shane Ellison, who has anointed himself as "The People's Chemist." Needless to say, "chemical-free" is a dually absurd expression. For one, nothing, save a vacuum, is chemical-free, since chemicals are just the building blocks of all matter. Secondly, the message implies that "chemicals" are synonymous with "toxins" or "poisons." Nonsense! Chemicals are not good or bad; their specific use has to be judged on their merits as determined through proper scientific investigation. How the words "chemical-free" can come out of the mouth of someone with a Master's degree in organic chemistry is a true mystery.

While "chemical-free" can be construed to be benign gibberish, urging people to "ditch your meds" is anything but. Ellison, who once worked for a pharmaceutical company, now tells people that the time has come to give up their blood pressure, thyroid, cholesterol-lowering, and antidepressant meds because these are evil concoctions foisted on the world by a heartless, profit-driven medical-industrial complex. Incredibly, casting insulin aside is also included in this deplorable, foolhardy advice.

Ellison's vengeful attacks also target COVID vaccines that according to him contain a "chemical s***storm to fight the common cold dressed up as COVID-19." He even links the vaccines to the production of Zyklon B, the notorious cyanide-releasing chemical used in the Nazi gas chambers. Zyklon B was made by IG Farben, a company Ellison says "after the war broke off into Bayer and Merck which became Moderna." That is ludicrous and even the history is wrong. Merck was around long before IG Farben, and Moderna is not a Merck spin-off. Just another example of the lengths to which rogues will go to gain a following. Not surprisingly, Ellison also claims that the "HIV/AIDS hypothesis was one hell of a mistake" and that "Dr. Fauci and other Pharma-fueled scientists" made millions from selling treatments that were unnecessary. Just totally mindless blather.

By now you have probably guessed that this sage offers an alternative to the drugs that he says should be ditched. A bevy of "chemical-free" supplements is available for purchase through his website. For example, after dissing vaccines and claiming they are not needed for a virus that doesn't exist, he pushes his own Immune FX, which contains an extract of two plants, Andrographis paniculata and coriander. Apparently, plants do not contain chemicals. A literature search for the medicinal value of these plants comes up with some evidence of coriander having antibiotic effects in the gut of broiler chicks, and some Andrographis compounds having anti-inflammatory and antioxidant activity when tested in cell cultures. There is zero evidence of Immune FX having been put to a test in human trials.

A supplement labeled as Serotonin FX suggests that it boosts levels of serotonin, a neurotransmitter that can indeed have an antidepressant effect as clearly demonstrated by the use of prescription selective serotonin reuptake inhibitors (SSRIs). Ellison's formula contains the amino acid L-tryptophan that, once again, is not considered to be a chemical. While L-tryptophan is indeed the body's precursor to serotonin, there is no clinical evidence of equivalence to SSRIs. Patients who have been prescribed these drugs are wading into deep waters should they replace them with Serotonin FX.

The other supplements hyped by the People's Chemist are similarly devoid of evidence. His answer to pain is Relief FX, with ingredients extracted from white willow bark and ginger root. The willow bark contains salicylic acid, which has pain-relieving properties but also irritates the stomach. That is why it has been replaced by acetylsalicylic acid, better known as aspirin. Synthetic aspirin is superior to natural salicylic acid, so going back to willow bark makes no sense. Ginger is a chemically complex mixture of dozens of compounds, some of which do have analgesic effects, but the label on Relief FX yields no information about what ginger-derived compounds are present or in what dosage.

Men who are concerned about testosterone levels are offered Raw-T, an extract of sarsaparilla root. But they will be disappointed if they replace their prescription testosterone with this concoction. Testosterone is a steroid, and sarsaparilla does contain related compounds called sterols; however, these are not converted in the body to testosterone. Neither is there any evidence that Preworkout, advertised as "Faster Higher Stronger in 59 Minutes Without Chemicals," delivers the goods. Never mind that citrulline, tyrosine, Hawthorne extract, yerba mate, and huperzine A are absurdly described as not being chemicals; the claim that any substance can make someone faster, higher, or stronger in 59 minutes sticks in the craw.

"The People's Chemist" curiously uses a microscope as his logo, an instrument not commonly used by chemists. He also describes how he walked away from an "award-winning career as a medicinal chemist" when he discovered that people taking cholesterol meds, cancer drugs, and blood thinners were just victims of Pharma's insidious marketing practices. Indeed, there are some reprehensible practices, and they should be brought to light, but that does not mean that all drugs should be feathered and tarred with blatantly farcical arguments. Ellison says that he poured blood, sweat, tears, and years into becoming a chemist. He could also have used a bit of sense poured into his head. With an attempt at humor, he adds that actually, there were no tears, because chemists don't cry. Wrong! His prattle about a "chemical-free" lifestyle can bring a tear to any chemist's eye. The only question is whether it is from laughing or crying.

ALTERNATIVE MEDICINE

"Alternative medicine" is a perplexing term. What does it mean? Medicine either works, or it doesn't. If it works, it isn't alternative. If it doesn't work, it isn't medicine. So what then is alternative medicine? The best definition seems to be "those practices which are not taught in conventional medical schools." And why not? Because medical schools are sticklers for a little detail called "evidence." After all, patients have a right to expect that a course of action recommended by a physician has a reasonable chance of working. In science, evidence means statistically significant results from properly controlled experiments, as evaluated by experts in the field. Lack of evidence, of course, does not mean that a particular treatment cannot work. Only that it has not been demonstrated to work. And that is when it can be termed "alternative." If sufficient proof is mustered, alternative transforms into conventional.

Today, the conventional treatment of ulcers often involves the use of antibiotics. That's because there is now clear-cut evidence that many ulcers are caused by the *Helicobacter pylori* bacterium. When the bacterial connection was first suggested by Drs. Barry Marshall and Robin Warren back in the 1980s, it was certainly in the alternative realm. After all, physicians "knew" that ulcers were caused by stress and excess stomach acid. Skeptics, appropriately, wanted evidence before they jumped on the bandwagon. And it didn't take long for it to be provided.

In a cavalier and somewhat foolhardy fashion, Marshall drank a solution of *Helicobacter pylori* and developed a case of gastritis. No ulcer formed, but he took a heavy dose of antibiotics. The experiment worked and managed to stir the scientific community into action, and within a few years hundreds of papers were published on the subject. Controlled trials were carried out, and antibiotics were clearly shown to be an effective treatment for ulcers. Today, this is the preferred treatment and is taught in every conventional medical school. Although initially some physicians may have scoffed at the idea of ulcers being caused by bacteria, they were quickly won over by the evidence. Contrary to what is often claimed by alternative practitioners, physicians are not closed-minded about approaches they may not have learned about in medical school, they just would like to see some sort of evidence of efficacy before advocating them.

Alternative medicine, by the proposed definition, encompasses a vast array of treatments, ranging from the possibly useful but unproven, to the totally implausible. Homeopathy, one of the most popular alternative treatments, falls into the latter category. We have already encountered this pseudoscientific practice in our discussion of Oscillococcinum, the reputed homeopathic treatment for the flu.

Critics of homeopathy have been known to swallow entire bottles of homeopathic pills to make the point they contain nothing but sugar. But homeopaths are not disturbed by this demonstration because according to the tenets of homeopathy, increasing the dosage actually reduces the effect. So, the critics would face danger not by taking more pills, but by just licking one. Or, perhaps, they could overdose by staying away from the pills altogether.

We can safely say that homeopathic remedies pose no risk of side effects or of toxicity. Just try calling a poison control center to say that you accidentally took too many homeopathic pills. You'll get a response along the lines of "forget it" or "bogus product." But does this mean that homeopathy presents no risks? Not at all. There are several concerns.

Some homeopathic remedies may not actually be homeopathic. More seriously, some homeopaths offer pills for protection against malaria or radiation exposure. Others claim that they can treat cancer, with the most outrageous ones urging their victims to give up conventional treatment. Finally, there is the matter of Health Canada issuing a DIN-HM (Drug Identification Number-Homeopathic) to homeopathic products, implying to the consumer that these remedies have been shown to be safe and effective. Safe, yes. Effective, no.

Let's amplify. Marketers sometimes use the term "homeopathic" to describe products that are not at all homeopathic. A classic case is Zicam, sold as an intranasal homeopathic cold remedy until 2009 when the Food and Drug Administration advised that the product be avoided because of a risk of damage to the sense of smell. How can a homeopathic remedy do that? Simple. It was mislabeled. Zicam actually contained a significant amount of zinc gluconate. This, though, is not nearly as serious as recommending ridiculous malaria protection pills that contain no active ingredient to people traveling to areas where the disease is endemic.

And how about Homeopaths without Borders? I kid you not. Here is one of their gems: "With the onset of the rainy season in Haiti there will be a great need for remedies to treat dengue, malaria, cholera and other tropical diseases." Claiming that homeopathy can treat these diseases is criminal. Jeremy Sherr of Homeopathy for Health in Africa goes even further: "I know, as all homeopaths do, that you can just about cure AIDS in many cases." Nonsense, of course, and even disparaging to most homeopaths, who draw the line at claiming cures for serious diseases.

Perhaps the most reprehensible practitioners of homeopathy are those who prey upon desperate cancer victims. The following comes from the Wisconsin Institute of Nutrition, whatever that may be: "The important thing to know about cancer and choosing whether to use homeopathy or not is that surgery will not remove the disease. Most people will still opt for conventional treatment, so how can homeopathy be useful to them? They can take the appropriate remedy after surgery to prevent recurrence. For strict homeopathic thinking such a procedure is not optimum." Needless to say, there is zero evidence that sugar pills can prevent a recurrence of cancer.

Homeopaths are not ones to miss a marketing opportunity. Soon after the Fukushima nuclear power plant disaster in Japan, several offered remedies for either the treatment or prevention of radiation poisoning. Believe it or not, one of the suggested remedies was "X-ray." What is it? A sugar pill treated with a homeopathic dose of X-rays. I wonder what equipment is used to dilute X-rays.

The coronavirus has also given homeopaths a chance to spread their nonsense. In this case, the treatment is with arsenic trioxide, albeit diluted to the extent that the final solution contains nothing. Needless to say, there is no evidence that arsenic in any form can destroy the coronavirus, at least not without destroying the patient. Amazingly, this bit of quackery comes from the Indian Ministry of Ayurveda, Yoga, Naturopathy, Unani, Siddha, Sowa Rigpa, and Homeopathy (AYUSH). Yes, India has such a ministry. The advice of treating coronavirus with homeopathy or with various Ayurvedic herbs has been roundly lambasted by the Indian Medical Association, as it should be.

Homeopathy has always been challenged by scientists, but consumers are beginning to realize the delusion of dilution. In California, homeopathic manufacturer Boiron settled a $12 million class-action lawsuit that alleged the company had violated false-advertising laws by claiming that homeopathic remedies have active ingredients. Boiron will now be adding a disclaimer to say that their claims have not been evaluated by the U.S. Food and Drug Administration as well as an explanation of how their active ingredients have been diluted. In Australia, a woman is suing a homeopath who she claims offered misleading information to convince her sister to give up conventional cancer treatment.

In Britain, the House of Commons Science and Technology Committee released a report stating that homeopathic remedies work no better than placebos and should no longer be paid for by the U.K. Health Service. The committee also criticized homeopathic companies for failing to inform the public that their products are "sugar pills containing no active ingredients." And at a British Medical Association conference, an overwhelming vote supported a ban on any funding of homeopathic remedies, calling them "witchcraft."

In Canada, the Natural and Non-Prescription Health Products Directorate has a mandate "to ensure that Canadians have ready access to natural health products that are safe, effective and of high quality." Yet, it licenses homeopathic products without requiring proof of efficacy. Why should the manufacturers of these products be less accountable than those of other pharmaceuticals? Knowing this, how can pharmacists in good conscience sell sugar pills that claim to have ghostly images of molecules?

Homeopathic remedies work through the placebo effect. That of course is not negligible. Placebos can have success rates of over 30 percent! But if you think there's something more to homeopathy, consider the following. How come different homeopaths prescribe different remedies to the same person for the same condition? How come drugs, other than homeopathic remedies, do not increase in potency when they are diluted? How come trace impurities in the sugar used to make the tablets, or in the water or alcohol used for dilution, which are present at higher concentration than the supposed active ingredient, have no effect? How can remedies that are chemically indistinguishable from each other have different effects? And how come a producer of homeopathic remedies given an unidentified pill cannot determine the original substance used to make the dilution? Finally, how come there are no homeopathic pills for diabetes, hypertension, or birth control?

Homeopaths at least do not usually claim to treat cancer. But there are plenty of alternative practices that claim success where conventional medicine fails. Claims of cancer cures are of course nothing new. In medieval Europe a live crab would be placed on the body at a site close to a tumor, left there for a while, and then the animal would be removed and killed. Why? Because many tumors were seen to bear a physical resemblance to the crab. In fact, our word cancer derives from the Latin word for the creature.

The idea, then, was that the tumor would develop some kind of association with the crab and would somehow be sympathetically destroyed along with the poor crustacean. Judging by the fact that this procedure persisted for a couple of centuries, it must have produced at least some successes. This is not surprising in light of our current knowledge about spontaneous remissions and the placebo effect. But even with the popularity of "alternative medicine" today, it is safe to say that anyone suggesting that crabs can physically withdraw cancer from the body would be regarded as less than sane.

Now fast-forward to the present. Imagine that you were suffering from cancer. Imagine that you were told that you could be cured of the disease in just five days by identifying and then removing the cause of your cancer. Imagine that all you had to do was buy about $35 worth of parts and build a simple electronic device that would tell you exactly what to do. Imagine that you were instructed to eat a certain food, then squeeze a pimple on your body and place the emerging fluid on the device next to a sealed plastic bag of the same food. Imagine that you were then to connect the contraption to your knuckles by means of two leads and listen to the sound emanating from a little speaker in the apparatus.

Now imagine that by the type of sound emitted you could determine whether this particular food was a cause of your cancer and must therefore be eliminated from the diet to ensure a cure. Finally, imagine that you don't have to imagine all this. For indeed, the foregoing is the actual scenario being plied to the public in an epic work with the grandiose title *The Cure for All Cancers*!

Hulda Regehr Clark, who surprisingly graduated with a PhD in physiology from the University of Minnesota, claimed to have discovered the secret that has stymied all other scientists. The cause of cancer, she proposed in her epic, is an intestinal parasite that can escape from the gut and take up residence in a variety of organs that have been weakened by previous exposure to substances ranging from mercury in dental fillings and thallium in wheelchairs to wallpaper glue and asbestos in clothes dryers.

But the cancer process can only begin if certain other chemicals are concurrently present in the body. Apparently the greatest culprit is isopropanol, otherwise known as rubbing alcohol. But other solvents such as methanol or xylene can also initiate cancer when present together with the parasite. These solvents, according to Clark, are found as contaminants in our foods, drinks, and cosmetics. The cure for cancer then is obvious to the writer. Kill the parasites and avoid all products contaminated with solvents as well as all chemicals that weaken our organs. These products include shampoos, cold cereals, carpets, stainless steel, porcelain, and toast. Toast, you ask? Of course. It is contaminated with tungsten from the element in the toaster. So we are told.

How does one go about killing the parasites? A mixture of cloves, black walnut, and wormwood destroys the intestinal flukes, as they are called, and therefore in Clark's words, "can cure all cancers." And of course the instrument just described, which Clark calls a Syncrometer, will determine exactly which foods and other substances must be avoided to effect a cure. If you want to know whether there is any aluminum in your brain, weakening it and therefore making it more susceptible to disease, the Syncrometer can tell you. According to the detailed instructions, just buy a piece of pork brain, place it on the device next to a piece of aluminum, attach the leads, and listen for "resonance." The pork brain, you see, guides the instrument where to look, and the piece of aluminum tells it what to look for. Similarly, you can use a piece of fish intestine to test for parasites in your colon.

How anyone can come up with such a bizarre concept boggles the rational mind. The story would be funny, if the possible consequences were not so sad. Hulda Clark's followers actually use her Syncrometer to diagnose cancer! They then go on to cure people of a disease they never had.

Clark, in one of many "case histories," describes how a patient had undergone a colonoscopy for severe diarrhea and had been pronounced cancer-free by her physician. Yet one of Clark's bizarre tests showed a positive reaction for cancer. "It came as a shock to her that she actually had colon cancer," Clark says. I bet it did. Of course, a week after starting on the anti-parasite program she was pronounced cancer-free. Strangely, the diarrhea was still present. One also wonders how many people who really may have serious disease resort to this "therapy" at the expense of proven remedies.

But Clark, in her book, is not completely anti-establishment. She does admit that oncologists are kind, sensitive, compassionate people. But "they have no way of knowing about the true cause of cancer since it has not been published for them. I chose to publish it for you first so that it would come to your attention faster." And publish it she did. Our doctrine of freedom of speech guarantees her right to do so. Unfortunately the doctrine does not require that what is stated be scientifically valid. Free speech emerging from the wrong mouth can be very dangerous!

And where is Hulda Clark today? She is no longer dispensing any form of alternative treatment. She passed away in 2009. From cancer.

SOME VIEWS ON DEALING WITH INFORMATION AND MISINFORMATION

1. Science is a process used to search for the truth. It is not a collection of unalterable "truths." It is, however, a self-correcting discipline. Such corrections may take a long time; bloodletting went on for centuries before its futility was realized. But as more scientific knowledge accumulates, the chance of making substantial errors decreases.

2. Certainty is elusive in science and it is often hard to give categorical "yes" or "no" answers to many questions. To determine if bottled water is preferable to tap water, for example, one would have to design a lifelong study of two large groups of people whose lifestyle was similar in all respects except for the type of water they consumed. This is virtually undoable. We therefore often have to rely on less direct evidence and educated guesswork for our conclusions.

3. It may not be possible to predict all consequences of an action, no matter how much research has been done. When chlorofluorocarbons (CFCs) were introduced as refrigerants, no one could have predicted that thirty years later they would have an impact on the ozone layer. If something undesirable happens, it is not necessarily because someone has been negligent.

4. Any new finding should be examined with skepticism. A skeptic is not a person who is unwilling to believe anything. A skeptic, however, requires scientific proof and does not swallow information uncritically.

5. No major lifestyle changes should be made on the basis of any one study. Results should be independently confirmed by others. Keep in mind that science does not proceed by "miracle breakthroughs" or "giant leaps." It plods along with many small steps, slowly building towards a consensus opinion.

6. Studies have to be carefully interpreted by experts in the field. An association of two variables does not necessarily imply cause and effect. As an extreme example, consider the strong association between breast cancer and the wearing of skirts. Obviously, the wearing of skirts does not cause the disease. Scientists, however, sometimes show a fascinating aptitude for coming up with inappropriate rationalizations for their pet theories.

7. Repeating a false notion often does not make it true. Many people are convinced that sugar causes hyperactivity in children — not because they have examined studies to this effect but because they have heard that this is so. In fact, studies have not demonstrated that sugar causes hyperactivity.

8. Nonsensical lingo can sound very scientific. An ad for a type of algae states that "the molecular structure of chlorophyll is almost the same as that of hemoglobin, which is responsible for carrying oxygen throughout the body. Oxygen is the prime nutrient and chlorophyll is the central molecule for increasing oxygen available to your system." This is nonsense. Chlorophyll does not transport oxygen in the blood.

9. There will often be legitimate, opposing views on scientific issues. But the impression that science cannot be trusted because "for every study there is an equal and opposite study" is incorrect. It is always important to examine who carried out a study, how well it was designed, if anyone stood to gain financially from the results, and where consensus lies. One must be mindful of who is the "they" in "they say that . . ." In many cases what "they say" is only gossip, inaccurately reported.

10. Humans are biochemically unique. Not everyone exposed to a cold virus will develop a cold. Response to medications can be dramatically different. Eating fish can be healthy for many but deadly to those with an allergy.

11. Animal studies are not necessarily relevant to humans, although they may provide much valuable information. Penicillin, for example, is safe for humans but is toxic to guinea pigs. Rats do not require vitamin C as a dietary nutrient, but humans of course do. Feeding high doses of a suspected toxin to test animals over a short term may not accurately reflect the effect on humans exposed to tiny doses over the long term.

12. Only the dose makes the poison, only the dose makes the cure. It does not make sense to talk about the effect of substances on the body without talking about amounts. In science, numbers matter! Licking an aspirin tablet will do nothing for a headache but swallowing two tablets will make the headache go away. Swallowing a whole bottle of pills will make the patient go away.

13. "Chemical" is not a dirty word and "chemical-free" is a nonsensical term. Chemicals are the building blocks of our world and are not good or bad; however, they can be used in safe or dangerous

ways. Nitroglycerine can alleviate the pain of angina or blow up a building. The choice is ours. Furthermore, there is no relation between the risk posed by a substance and the complexity of its name. Dihydrogen monoxide, after all, is just water.

14. Nature is not benign. The deadliest toxins known, such as ricin from castor beans or botulin from the Clostridium botulinum bacterium, are perfectly natural. "Natural" does not equate to safe, and "synthetic" does not mean dangerous. The properties of any substance are determined by its molecular structure, not by whether it was synthesized in the laboratory by a chemist or by nature in a plant.

15. Perceived risks are often different from real risks. Food poisoning from microbial contamination is a far greater health risk than trace pesticide residues on fruits and vegetables.

16. The human body is incredibly complex and our health is determined by a large number of variables that include genetics, diet, the mother's diet during pregnancy, stress, level of exercise, exposure to microbes, exposure to occupational hazards, and luck!

17. While diet does play a role in health, the effectiveness of specific foods or nutrients in the treatment of diseases is usually overstated. Individual foods are not good or bad, although overall diets can be described as such. The greater the variety of food consumed, the smaller the chance that important nutrients will be lacking in the diet. There is universal agreement among scientists that increased consumption of fruits and vegetables is beneficial.

18. The mind-body connection is an extremely important one. About 30–40 percent of people will improve significantly when given a

placebo and about the same percentage will exhibit symptoms in response to a substance they perceive as dangerous. The mind is capable of making a heaven of hell and a hell of heaven.

19. Roughly 80 percent of all illnesses are self-limiting and will resolve almost no matter what kind of treatment is being followed. Often a remedy receives undeserved credit. Anecdotal evidence is unreliable because positive results are much more likely to be reported than negative ones.

20. There are no geese that lay golden eggs. In other words, if something sounds too good to be true, it probably is. As H.L. Mencken said, "Every complex problem has a solution that is simple, direct, plausible, and wrong."

21. Education is not a vaccine against folly. An MD or PhD degree does not confer immunity to nonsensical beliefs.

22. Life is full of risks, but risks have to be weighed against benefits.

23. "Alternative" medicine is a misnomer. Medicine either works or it doesn't. If it works, it isn't alternative. If it doesn't, it isn't medicine.

24. The best source of scientific information is the peer-reviewed literature, but it is not infallible. Scientists do make errors and fraud is not unknown.

25. Nobody has a monopoly on being right, so don't guide your life by any set of rules produced by any individual. As Will Rogers said, "Everybody is ignorant, only on different issues."

INDEX